CRYPTO USERS' HANDBOOK

CRYPTO USERS' HANDBOOK

A Guide for Implementors of Cryptographic Protection in Computer Systems

Prepared under the auspices of IFIP WG 11.4

Per CHRISTOFFERSSON
Stig-Arne EKHALL
Viiveke FÅK *(editor)*
Siegfried HERDA
Pakka MATTILA
Wyn PRICE
Kjell-Ove WIDMAN

1988

NORTH-HOLLAND
AMSTERDAM • NEW YORK • OXFORD • TOKYO

ISBN: 0 444 70484 1

Publishers:

ELSEVIER SCIENCE PUBLISHERS B.V.
P.O. Box 1991
1000 BZ Amsterdam
The Netherlands

Sole distributors for the U.S.A. and Canada:

ELSEVIER SCIENCE PUBLISHING COMPANY, INC.
52 Vanderbilt Avenue
New York, N.Y. 10017
U.S.A.

PRINTED IN THE NETHERLANDS

FOREWORD

This book is the result of a project within IFIP WG 11:4. Its authors are the members of that Working Group. It has been made possible through the support of the employers of these WG members.

Its intended level is between a sales brochure or a short introduction on one side and a full text-book on the other. It is rather technical in nature, but tries to treat the subject without unnecessary technical vocabulary. The reader is supposed to have a reasonable knowledge about computers and computer security problems in general. Some sections, like the section on examples of encryption in communications protocols or the one on algorithms, go fairly deep into technical details, but explain the specific vocabulary in the text. There is also a glossary at the end of the book.

The text is rather condensed and full of facts and explanations. Still it stops at a certain level of detail. For further reading there is a list of references as the last item in the book. The entries there in turn contain references to even more detailed and specialized texts. Thus this book is meant to be sufficient for those who want to learn about cryptographic protection in a computer environment without specializing in the field. It is also meant to be a good introduction to those who later want to learn more.

Authors:

Per Christoffersson (Chapter 5)
Tele-K AB
Sweden

Stig-Arne Ekhall (Chapter 4.2)
Fst/TSA
Sweden

Viiveke Fåk (Chapter 7, editor)
Linköping University
Sweden

Siegfried Herda (Chapter 2)
Gesellschaft für Mathematik
und Datenverarbeitung mbH
Federal Republic of Germany

Pekka Mattila (Chapter 1)
OKO BANK
Finland

Wyn Price (Chapter 4.1 and 6)
National Physical Laboratory
United Kingdom

Kjell-Ove Widman (Chapter 3)
Crypto AG
Switzerland

CONTENTS

1 INTRODUCTION

If you and your company have nothing to protect, if all the information you are processing can be published, if your product development is not pioneering, if distortion of the information is not a threat to your company, if you are not competing, and if you do not use computers we thank you for buying this book, but do not bother to read it, because you do not need cryptography.

If you instead process data which contain meaningful information and if the distortion or disclosure of that information is a threat, you should utilize cryptography.

Computerized cryptographic techniques have been used for military purposes for decades, but in the commercial environment cryptography has not become as common as might be expected. The reason for that is that the technique has been too expensive compared to the need for protection. However, the situation is clearly changing. Today the crypto hardware and software has become less expensive, and the need can be seen more obviously. In fact cryptography has become an everyday tool also in commercial systems.

Information is a very important resource of a company. You can invest money in it and expect it to bring a good return. You must be able to trust the quality of the infomation and to be convinced that confidential or secret information cannot fall into the hands of an outsider.

There are risks in processing, storing, and transferring data. The data can be distorted or falsified during processing. The registered data can be copied or changed without authorization. While transferring data via tapes, diskettes, or communication network they can be directed to false hands or even forged.

What sort of problems can you solve with cryptographic methods? Cryptography is the only practical means for protecting the transfer of data, and in many cases it is also the best way to protect stored data. Cryptography is the process of transforming information (cleartext) into an unintelligible form (ciphertext) so that it may be sent over insecure channels or it may be stored in insecure files. Today, at the time of open networks, hackers and crackers, we cannot consider any communication channel or file storage to be secure as a default.

With encryption (enciphering, scrambling, coding) you can translate clear data to a form which an outsider cannot interpret unless he or she knows the method and the possible key used in the process.

Decryption (deciphering, unscrambling, decoding) is the inverse process in which encrypted data are translated to clear data. By encrypting the data before transferring or storing them and by decrypting them again before processing you can make sure of their confidentiality.

The company can make vital decisions based on the information printed by computers. The forged data look as "genuine" as the original. Therefore the authentication of data must be self-evident when you handle information which will affect the decisions. With cryptographic techniques you can discover the distortion or falsification of data whether they are money transactions, strategic budgeting information or even program modules.

You must realize that cryptography can neither prevent an outsider from obtaining secret data nor can it stop anyone from falsifying transmitted or stored data. But data **are not** information. Data just **contain** information. Having encrypted data frustrates an outsider, because he or she cannot utilize it. Changing authenticated data brings no profit, because the receiver does not process unauthorized changed data.

Cryptographic procedures can also be used for personal identification, digital signature, access control, etc. The crypto applications are discussed in chapter 2 in greater detail.

The crypto system usually has a cryptographic algorithm, which may be secret or public, and a character string, which is called the key. In most solutions the key and its secrecy are the basis of the whole system. The measures by which the keys are generated, stored, transferred, and changed are called key management. In most cases it refers to a very complex hierarchic system.

Very often thorough key management is neglected. However, you must remember that the key management can be the weakest link which may break the whole expensive system. The algorithms and key management are described in chapters 3 and 4 respectively.

When you decide to use cryptography you must remember two things: 1) the decision is not enough and 2) the encryption is not enough.

Cryptography is not the easiest way to protect data. There will be many obstacles in the way. The speed of an algorithm can be a restriction in transaction processing; the complexity of the network can make the solution too expensive; the creation, storage, and distribution of keys can be the most difficult task. If you decide to use cryptography, you should plan the project very carefully. A crypto system which is carried out poorly gives only a false sense of security but does not protect the information.

Cryptography is a powerful tool to protect the secrecy and integrity of information, but it cannot alone make your whole data processing environment secure. To achieve an acceptable level of security you must also use other appropriate security precautions.

2 APPLICATIONS

Applying cryptographic techniques to meet security requirements one has to keep in mind that the measures to be selected must remain manageable by all parties involved, especially by the user, who, as a rule, knows little about cryptographic techniques. The danger of unmanageability may arise from various sources. The first is the fact that encryption keys have to be distributed to all places where the cryptographic transformation is to be performed. This key distribution problem will be discussed in chapter 4.1 as part of the key management issue. Another reason is the fact that cryptographic techniques are often not properly embedded into the environement because off-the-shelf products sometimes do not exactly meet the particular requirements of a specific security problem. Another drawback may arise from the fact that standardization is lagging behind the development causing potential incompatibilities among products of different manufacturers.

However, the situation has been changed during the last few years with the advent of public-key cryptosystems, and by the fact that standardisation has become a world wide (ISO and CCITT) joint effort, and is kept in step or even ahead of the development of products. This new situation reduces the danger of cryptographic techniques tending to become unmanageable. These developments, and especially public-key cryptosystems, offer new application areas, or make possible more elegant solutions to old problems. These developments in cryptography are supported by other technological developments such as the chip card technology. This promising situation reduces the risk that may arise if cryptographic techniques are not properly implemented and embedded.

In the following paragraphs we give an overview of how cryptographic techniques can be applied starting with the classical field of protecting information against unauthorized reading of the data in transmission and in storage, the problem of confidentiality. Then, authentication of messages and of their origin is described using old (authenicators) and new (digital signatures) technique.

The problem of having public keys certified by some agreed authority is touched upon by discussing the need for a trusted third party (or authentication server). Techniques to verify the identity of an entity whenever claimed are described in section 2.3. An outlook on software protection and how it can be provided by cryptographic techniques will finish this chapter.

2.1 Confidentiality

Hiding confidential information has a very long tradition. It is not surprising that binary representation of characters in electronic computers stimulated the development of new cryptographic algorithms. Cryptographic techniques based on these algorithms are much stronger than traditional techniqes. But powerful cryptanalytic techniques have been developed at the same time. Cryptographic transformations of data for secrecy reasons are sometimes called privacy transformations.

Cryptographic techniques are applied to data in transmission or in storage. The questions to be answered are where to place the techniques without impairing the system, what should be enciphered, is the key management problem manageable? Data are often stored in a computer primarily for use by more than one person. Is it then possible to apply encryption on a selective individual basis? Do enciphered data allow users to perform statistical operations on a database?

Enciphering data in transmission

Encipherment can be employed in three ways: on the "line level", "end-to-end", and "link level".

Implementing encipherment on the line level offers a high degree of protection. Everything passing over that line is enciphered regardless of where the data originate. A wiretapper is unable to analyse the traffic. The security depends on the cryptographic equipment selected. The enciphering facility has to be provided by the operator of that line. This may be the company that leased the line, or the national PTT. Key management is rather simple. Only the two points connected by the line are concerned. If the line connects nodes of a network and not two "end-users" the information in the nodes is in clear. This is a big disadvantage of line level encipherment.

This disadvantage can be avoided by end-to-end encipherment. With end-to-end encipherment the data remain encrypted until they reach their final destination. End-to-end encipherment is often required for electronic mail, and for electronic funds transfer. However, end-to-end encipherment can be susceptible to traffic analysis. Key management can be a problem if users need different keys for communicating with each correspondent. The communicating users must of course use the same encipherment algorithm.

These drawbacks are avoided using the hybrid way of enciphering data in transmission by link encipherement, also known as node-by-node encipherement. It may be regarded as end-to-end encipherment between every two nodes of a network connected to each other. Each node has a unique key for each link. That means each incoming message is deciphered and enciphered again within the node (see Fig. 2.1). Link encipherment relieves the end-user of the burden of key management. Key management is fairly simple for the network operator. However, routing information is transmitted in clear so that traffic analysis is possible. Nodes become sensitive points with this method.

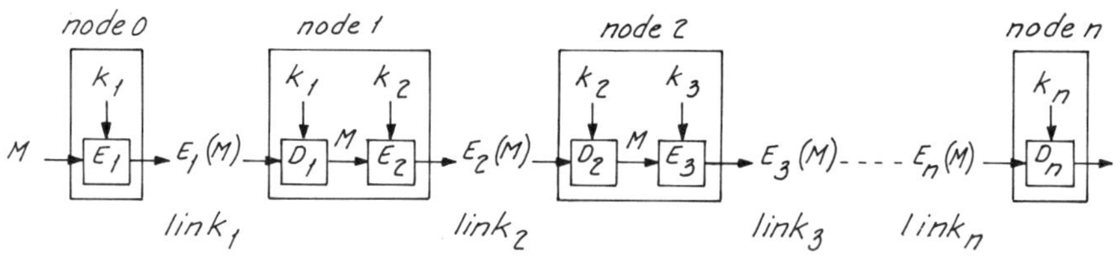

Figure 2.1 Link Encipherment

The highest level of protection is obtained by combining either line level encipherment or link encipherment with end-to-end encipherment. Line level and link encipherment is the task of the network operator, the end-to-end encipherment is the task of the user. Therefore, implementation of different crypto-devices and of different key management schemes is separated.

Enciphering data in storage

In most mainframe computers data on external storage, such as files and databases, are protected by access control mechanisms based on more or less sophisticated authorization schemes. Users are authorized to access the data by attributing to them appropriate access privileges, and/or passwords. This provides only limited protection against intruders and hardly any protection against privileged users, e.g. system programmers, operators, who can bypass the access control mechanisms. Physical protection of storage media, such as disks, magnetic tapes, diskettes, etc. is also limited. They may be stolen or illicitly copied, and thus sensitive information may be disclosed to unauthorized persons. Encipherment of data on storage media is a counter measure for such security violations.

The level on which encipherment is applied depends on the sensitivity and usage of the data and on the overall security policy of the installation. Encipherment of entire files is used where the content is not structured into records and fields, such as text files, source programs, or where the file is shared by only a group of users with the same privileges. Most applications require protected access to the record and/or field level of a file. Then each record and/or each subfield (or block) of all records must be enciphered separately. The most general place for stored data to be shared among a large user population is the database. Encipherment techniques for databases must be designed and implemented into the database management system (DBMS) by the database manufacturer.

There are a number of problems around the enciphering of data in storage compared with enciphering of data for transmission:

* The time available for a cryptanalyst is only limited by a key change (re-encipherment), or by the lifetime of the data.

* The data may also exist in clear, e.g. in main memory used by another user, giving the analyst the opportunity to perform a known-plaintext-ciphertext attack.

* Subfields are often smaller than the block length of most encipherment algorithms, so that other modes of operations must be applied.

* The demand to perform arithmetic operations on enciphered data directly (without deciphering) requires special encipherment algorithms, or modification of the query processor.

* The speed of the I/O devices demand adequate speed of the cipher device. That may require a much faster encipherment algorithm, and/or its implementation in hardware.

* Enciphered data in storage require a safe storage for keys that may be located in the central computer, in the intelligent terminal, or on a chip-card.

The distribution of keys to access enciphered data on various levels (file, records, subfields) may be regarded as a part of the authorization scheme. In user controlled file encryption the owner may select a key and distribute it to other users. That key may be stored in the system protected with a system key (or master key) applying the scheme described in chapters 4.1 and 5.2. The enciphered key(s) must be protected against modification. For keys with a long lifetime it is highly recommended to have them stored in firmware.

If the records of a file are structured into subfields and thus each subfield is enciphered with a subkey, key mangement becomes crucial. One elegant solution has been proposed by Davida, Wells and Kam (see References, section 2). They introduce a special block cipher algorithm for databases modelled as a set of records divided into subfields. In this scheme each record is enciphered as one unit so that all fields are diffused over the ciphertext of the record. Access to a particular field requires the subkey for that field, i.e. the individual fields can be separately deciphered by retrieving the entire cipertext record.

A variant of this scheme allows the computation of statistics, such as averages, standard deviations, etc, using only enciphered data. Detailed discussions of key management schemes are given in chapter 4.1. A quite different approach of processing data in storage is to implement the encipherment facilities in the central I/O unit or in the I/O controller.The keys are totally under system control. From the user's point of view encipherment/ decipherment is done automatically. This apporach protects the contents of peripheral storage media from being disclosed in case they are stolen.

2.2 Data Authentication

Protection against passive attacks, such as eavesdropping, is ensured by enciphering data for secrecy. There is also a growing need to protect data against active attacks, such as alteration, deletion, addition/insertion, or changing the origin and/or destination, using previously transmitted or stored data (replay), or falsifying acknowledgements. Autentication is the technique to counter these active attacks. Communicating parties not only want to ensure that the message transmitted has not been falsified, they want also to be sure that the message originates from the entity that claims to be the originator. In the first case there is the problem to ensure message authenticity, in the latter the problem to ensure the identity of an active entity, i.e. to ensure the authenticity of communicating parties (peer entities), or the authenticity of the originator of a document. Moreover, in transactions of commercial or legal value the authenticity of a whole session must not be impaired in any step. This is the problem to maintain session authenticity, also called session integrity.

Authentication is the procedure for ensuring the authenticity in any of these cases. Authenticity is established by communicating authenticators, i.e. a piece of data derived from the message, from the information belonging to or characterizing an active entity (user, originator), or from previous steps of relevance in a session. More precisely, authentication consists of two procedures, one for generating the authenticator, and one for verifying the authenticator.

Enciphering data for secrecy reasons protects only against insertion of new data, but not against insertion of ciphertext intercepted from previous transmissions and generated with the same enciphering key. Messages usually contain sufficient redundancy so that the receiver of the enciphered message gets a meaningfull message after deciphering it. This is regarded as the "self-authenticating property of encipherment". To detect the illicit insertion of previously enciphered data more effectively redundancy must be added to the data before enciphering them. This additional field is often called "manipulation detection code" (MDC). Procedures for a MDC are going to be standardized.

There are good practical reasons to separate the secrecy question from the authenticity question. In many applications authentication is required whereas secrecy may be of minor importance, or even a nuisance.

Methods for generating authenticators

1. Authenticators generated without a secret key

To authenticate a message M the authenticator A = f(M) is calculated using a "one-way function" f without secret parameters. A one-way function is easy to compute, but its inverse is computationally hard. The range of the value A must be large so that it is infeasible to find another message M' resulting in the same A, f(M') = f(M). However, everybody knowing the one-way function applied is able to forge the authenticator and the message unless the authenticator is communicated to the receiver or reader separately, or is protected by other means. We will refer to that problem in the next paragraph.

2. Authenticators generated with a secret key

"True" authenticators rely 1) on the message M, and 2) a secret parameter, a key k, known only to the sender and the receiver (or reader). The function f (k,M) may be public, and the message M may be of arbitary length. The authenticator A = f (k,M) must depend on every bit of the message M. Its length n depends on the function used, and is typically 16 to 32 bits, depending on the application area. The probability of finding another message with the same authenticator value is one in 2^n where n is the number of bits in A.

The authenticator is added to the message to be transmitted. The receiver calculates the authenticator A' = f (k,M') from the message M' received, and the secret key k. If both authenticators are equal, the authenticity of the message is ensured.

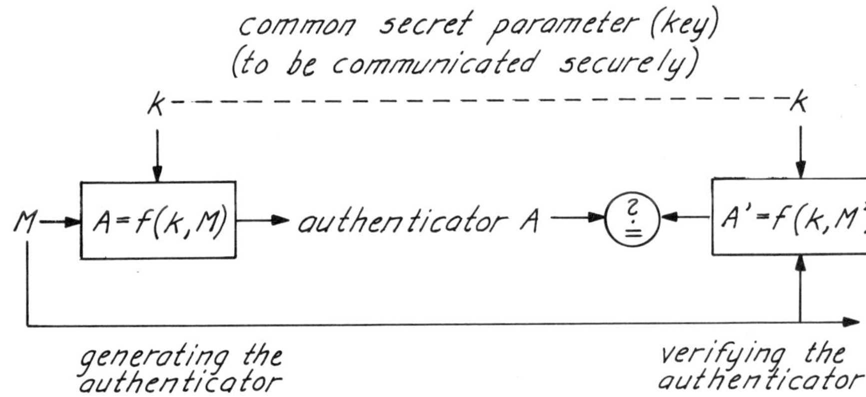

Figure 2.2 Principle of message authentication

Authenticators are widely used, especially in banking for transmitting financial messages. One example is the authenticator of the Society for Worldwide Interbank Financial Telecommunication system (S.W.I.F.T.) where the authentication function is kept secret for additional security.

Authentication algorithms have been proposed to ISO for standardisation. The Decimal Shift and Add (DSA) algorithm is based on decimal arithmetic, but was shown to have some weaknesses. A "Main Frame" authentication algorithm, suitable to run on computers, is based on 32-bit arithmetic and logical instructions, but its usage is not restricted to mainframes.

Authenticators based on standard "Modes of Operation"

If a hardware or a fast software implementation of a block cipher algorithm is available, it is convenient to use it for generating an authenticator. The "modes of operations" defined in FIPS Pub 81 for use of the DES may be used for any block cipher algorithm. There are two authentication methods, one based on the cipher feedback (CFB), the other on the cipher block chaining (CBC) mode generally used for enciphering messages. (See chapter 3.5 for a detailed description of chaining modes).

Generating an authenticator in the CBC mode, the message to be authenticated is divided into 64-bit blocks. An incomplete final block is padded with binary zeroes. Each block is enciphered and XORed (added modulo-2) to the next plaintext block to be enciphered. The most significant n bits of the last enciphered block are taken as the authenticator. All previous enciphered blocks are discarded. The number n of bits to be used is subject to an agreement of the communicating parties. For financial messages n is recommended to be 32, so that the probability of finding a substitute is reasonably small, $1/2^n$. In some countries the authenticator is called "message authentication code" (MAC).
In the banking area the authenticator is widely called "test key" causing confusion with the secret key used by the cipher algorithm to generate the authenticator.

Authentication of data in storage

The authentication techniques may also be applied to data in storage. Among other problems, there is the problem of comparatively long lifetimes of the keys. Changing the key requires a recalculation of the authenticators. Data authentication methods have been proposed for verifying the authenticity of programs or files on external storage media. The authenticator is stored in firmware accessible by the supervisor or data management system only. Each time a program is brought into main storage, an authenticator is generated "on-the-fly" and compared with that stored in a reliable form. It has been proposed to furnish certified compilers with an authenticator supplied by the certification authority. This authentication technique could be a countermeasure against computer viruses.

Limitations of authentication based on authenticators

Authentication schemes described so far are useful against external threats provided the knowledge of the secret information (one-way function, secret key) is confined to the parties generating and verifying the authenticator. However, these authentication schemes are justified for use mainly in closed user groups with a limited number of participants such as in the banking sector. Anybody forging authenticators is out of the club, probably without having the chance to return.

Generally, if it comes to dispute (e.g., before a court), nobody can decide who changed the data because both parties have the same common secret information, and consequently accuse each other of having changed the data (and the authenticator).

Moreover, any third party knowing the one-way function or the secret key is able to generate authenticators and to verify the authenticity of the data, and to some extent the authenticity of the originator.

A solution to this deficiency is offered by digital signatures as described in section 3.7. Digital signatures are based on public key algorithms. Their outstanding feature is that they have two keys, one for encrypting the data and one for decrypting, and that one key (the encryption key) can be publicly released. The other one, the decrypting key, has to be kept secret. Public key cryptosystems allowing first to "decrypt" a message and then to "encrypt" it afterwards are called signature schemes. The signature is generated by the consignee (or signer for short) by applying his secret key to the message and this signature can be verified by anybody applying the signer's public key. Signature schemes provide an excellent means to split up the secret information used for generating authenticators into secret information known only to the signer, and to public information virtually known to everybody (especially a judge) to verify the signature, i.e. to verify the authenticity of the message and that of the originator (signer).

Digital signatures are characterized as follows:

1) digital signatures depend on the message, and on
2) secret information known to the originator/signer only;
3) public information known to everybody may be used to verify the signature, and, thereby, to authenticate the signer (verify his identity) and the message (verify its authenticity).

It is obvious that digital signatures encompass and surpass the properties of authenticators described above. Therefore, digital signature techniques can be used instead of the authentication techniques mentioned.

Transforming a message into a signature does not provide secrecy. Anyone getting the signature is able to recover the plaintext by enciphering the signature with the publicly known public key of the signer. If both signature and secrecy is needed encipherment and signature tranformaiton can be combined. The most convenient way is to sign the message first with the signer's secret key, and enciphering the signature with the receiver's public key as shown in Figure 2.3.

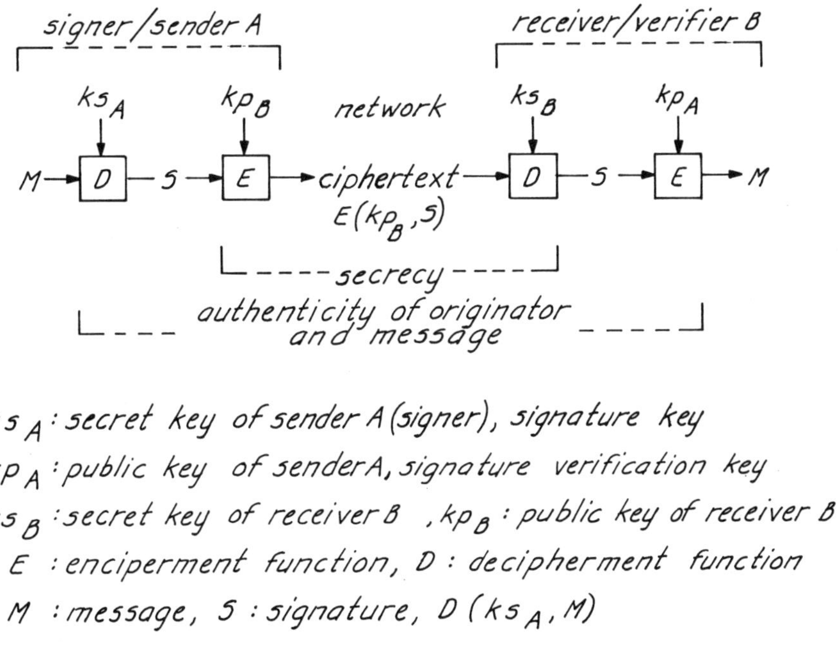

Figure 2.3 Public-key cryptosystems used for both signatures and secrecy

After receiving the ciphertext the receiver deciphers it with his secret key, and can store the recovered signature for future evidence. Taking the reverse sequence of transformations, i.e., first encipherment, then signature transformation, will cause some difficulties in presenting the signature for evidence to a judge.

There are two ways of deriving digital signatures from a message. The first one, described so far, is to decipher the whole message with the signer's secret key so that the resulting signature is at least as long as the plaintext message. The other way is to apply first a publicly known one-way function calculating a digest or hash depending on every bit of the message, and to sign that hash result. The signature is appended to the plaintext message. Signature verification is very convenient. The verifier (e.g., receiver) hashes the plaintext message, and compares this hash result with the result he gets from enciphering the signature with the signer's public key. This way of separating the signature from its message is becoming widely used.

2.3 Authentication of identity

Authentication of identity, or, synonymously, identity verification, is a difficult issue in the computer and communications environment. Most techniques are developed for identifying persons towards a computer based service. For mutual authentication more elaborate techniques are being developed.

User identification methods are based on some characteristics of humans. They are divided into four cathegories:

1. something the person knows, e.g. password, PIN,
2. something the person has (possesses), e.g. physical key,
3. physical characteristics of the person, e.g. voice,
4. something the person is able to perform, e.g. handwritten signature.

Asking for access to a computer, a communications facility, or to a service, such as cash dispensing terminals, automatic teller machines (ATM), or point of sale terminals (POS), one always needs to authenticate oneself using some secret information, usually a password or PIN (personal idetification number) issued by the provider of the facility. This password or PIN is checked at every access attempt after identifying information, a user-id, has been submitted.

Static passwords and PINs suffer from the danger of being intercepted and illegitimately used for impersonation. There are some one-way functions (mentioned in section 2.2) for translating passwords for storing, and at the time of reception the password is translated in the same way at the destination and then it is compared with that in storage. That avoids the risk of passwords being disclosed in clear when they are in main storage. They must of course also be encrypted during transmission. This different encryption may also be one-way. The crucial point of password based techniques is, however, that a secret is given into an insecure environment assuming that it remains a secret. Stronger identification schemes are based on cryptographic methods.

As mentioned at the end of section 2.2 digital signatures provide the ability to authenticate the identity of an entity by associating it with the secret key of the key pair. The problem, however, arises of how to connect that secret key to the entity so that it is impossible for anybody to disclose or copy that key. Otherwise the owner could claim that someone else generated the signature. Chip-card technology offers a means to store the secret key somewhere where it can neither be manipulated nor copied. The next step in chip-card development will be to furnish it with a keyboard and a display. That provides the additional capability for the owner to enter a PIN in order to authenticate himself towards the chip-card, and to activate it, e.g. to sign a message, without using equipment outside of his control. These chip-cards are called intelligent token, intelligent chip-card, or super smart-card.

In private and commercial life documents of relevance have to be signed, e.g., payments, contracts, etc. Even if the public key is taken from a key registry there remains the problem of trust, especially where many parties are affected by the signed document of legal or commercial value. Moreover, keys usually have a limited lifetime, and it is difficult to keep historical records of all keys of each key owner. Both problems are solved by introducing the concept of the "common point of trust" within a hierrachy of certification authorities. Some public information of the entity on each level of the hierarchy is certified by the next higher authority up to the highest node. The public key of that node is commonly known and a historical record of old keys can be kept without problems. A common point of trust is one that is either generally accepted, e.g., the highest node or a common certification authority in the tree of certification authorities of both the originator (signer) and the verifier of a signature. The public information to be certified consists of the distinguished name (may also be a pseudonym), the public key, the date of certification (time of issue), and, depending on the application area, the validity period of the cerificate. A ceritficate consists of this collection of public information signed by the ceritfication authority.

Examples of certification hierarchies are either: customer - customer's bank - central bank - national bank, or in the same manner: notary public, chamber of commerce, PTTs.

One or more of these cerification hierarchies may conveniently be stored in a safe form in an intellegent chip-card especially for use of payment of telematic sevices. In Figure 2.4 an example of a certification tree is given, developed for use of telematic sevices. It has been developed within the Authentication Framework that is part of the Directory System being standaridzed jointly by CCITT and ISO as CCITT Draft Recommendation X.509, and ISO DP 9594/8.

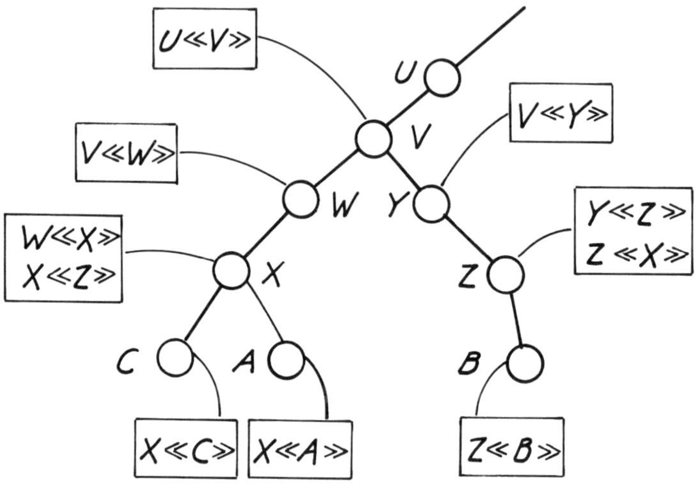

Figure 2.4 Certificates in the Certification Tree

Notation: X<<C>>: certificate of user C produced by the certification authority X

The certificates are stored in the directory system that will become availabe in the near future as a distributed data base from which customers may obtain the addressing and routing information as well as the certification path to any entity (customer) he wants to communicate with in a secure and authentic way.

Acess Authentication Protocols

The problem of authenticating communicating entities is discussed by presenting three authentication protocols. Which one to select depends on the level of protection required, or on the facilities available.

1. One-Way Protocol: After submitting the user-id a password (or PIN) is transmitted in clear without any cryptographic transformation. This establishes only a low level of security. If more intelligent equipment is available, e.g., an intelligent terminal, the password is encrypted and sent to the destination (the called party). This transformation may be one-way. In that case the destination computer compares the password submitted with that in its list. Where the password transformation is reversible, the computer decrypts the password submitted with a key that may be user specific, or time dependent. Another variation of this protocol is to derive the password or PIN from the user-id with a one-way function that is kept secret. This technique can be found in network access authorization. In networks there is always the problem of where to store the passwords and, if stored, how to fetch them in a reasonable amount of time.

2. Two-way Protocol: After being called from the calling party, the called party responds with a unique random number, unique in the sense that it is unpredictable and only used once. This random number is signed by the calling party with its secret key. This signature and its associated public key is sent back to the called party for verification. Before signing, a time stamp is added to the random number establishing a validation period to the signature to prevent replay. This two-way protocol is not secure against impersonation. A wire-tapper may record the reply message from the calling party, and may immediately transmit it to the called party. In a store and foreward application the attacker may have a chance to masquerade as being the true caller. In a two-way protocol the authentication is not mutual. Only the called party verifies the authenticity of the calling party. Mutual authentication of the communicating parties is established with the three-way protocol.

3. Three-way Protocol: The weakness of the two-way protocol is prevented by introducing another random number to the message sent to the called party. This random number is signed by the called party and returned to the caller in a third step.

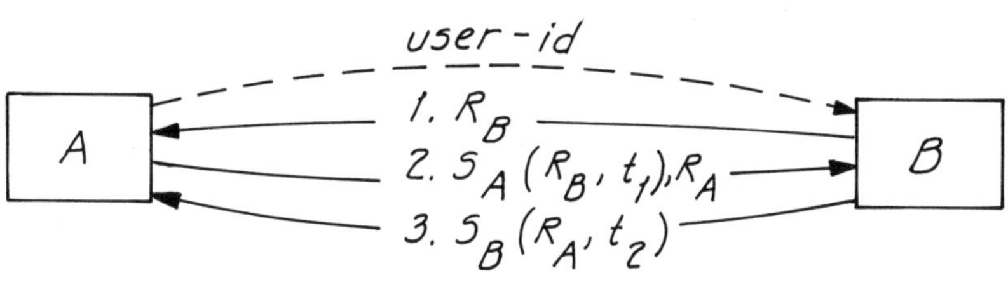

Figure 2.5 Access Authentication Protocol

This three-way authentication protocol does not establish the integrity of the whole session. Session authenticity can be maintained by chaining the signatures of each party in order to prove later on that these signatures belong to this particular session.

Identification Schemes

Signature schemes based on public-key algorithms with two keys where one can be made public and where the secret key can be uniquely associated with the owner, are very convenient for identification purposes as shown above. The need for a 'true' identification scheme has been claimed where the identifying information involved is only resident with the person to be identified. An identification scheme meeting this requirement has been developed by Fiat and Shamir, and is based on the zero-knowledge interactive proof-system (see References, section 2). A kind of secret key is derived from personal information describing the individual. The scheme is very efficient and highly interactive, and may primarily be used in online applications where communication costs are negligible, such as POS, ATM etc. The security of the scheme is still under scrutiny by the crypto-community.

Identified Transactions versus Anonymous Transactions

In transactions with organisations in the private and public sector there exists the potential danger for the individual that the identifying information is used to create a dossier containing information on the individuals' life-style, habits, whereabouts, and associations. Chaum developed an approach to avoid this assault on privacy allowing all parties involved to protect their own interests. It is based essentially on the concept of digital pseudonyms, and a facility to generate public-key pairs by each individual. The feasibility of this approach was demonstrated for three kinds of consumer transactions: communication transaction (e.g. electronic mail), payment transactions and credential transactions. The interested reader is referred to the texts by Chaum listed in References.

2.4 Software Protection

Protecting software from piracy is one of the most crucial issues in computer practice. Piracy has become more and more attractive as software costs are increasing whilst hardware costs are declining, and software is trivial to duplicate. Therefore, the development of large software systems has become increasingly risky. From the manufacturers' point of view a way to recoup their heavy investments and to protect their profit is urgently needed. Software protection has to address the problem of protection against illegal duplication and the problem of protection against illegal distribution. Protection against illegal duplication consists of maintaining that there is no cost-effective way for creating executable copies. Protection against illegal distribution consists of maintaining that, if duplication is possible, the unauthorized duplicator is unable to prove in court that he was entitled to distrubute copies. Preventing software from being illicitly duplicated helps to solve the problem of software being illicitly distributed.

Conventional methods of software protection, such as copyright laws, licensing contracts, non-disclosure contracts, etc., may be useful for mainframe computers, but have proven themselves to be ineffective for small computers, e.g. PCs.

Software protection methods currently in use only prevent unskilled users from creating copies, because many of them use software measures provided by the opeating system. Some others use the serial number of the computer, or a combination of serial number and information from the software, both put into a ROM that must be connected to the computer before the program can be executed.

In addition to providing insufficient protection, there are other deficiencies in many commercially available measures, such as impaired back-up capabilities, excessive overheads in cost and execution time, obstacles for authorized distribution and maintenance. The other extreme to software solutions would be to rely totally on hardware solutions, i.e. to sell physically protected special-purpose computers for each task. That approach is obviously infeasible.

It is agreed that effective software protection methods rely on some additional hardware in combination with cryptographic techniques. There are solutions suitable for all levels of protection depending on what is put into hardware, and on how the cryptographic protocols are designed.

The basic idea is to encipher the software to be protected so that it can only be deciphered and executed by presenting the appropriate key. The software protection problem is reduced to a key management problem, i.e. key generation, distribution and key protection.

Special hardware is necessary for the decipherment processor and for storing the keys. The enciphered software may be distribued by any distribution medium, such as diskettes, magnetic tapes, ROMs, or via telecommunications networks. To put some measures into hardware renders the overall system more complex. Not only the software manufacturer, the software distributor, the vendor (dealer), and the customer are involved, but also the computer manufacturer, and, most certainly, the OEMs. However, cryptographic techniques provide an excellent means for both solving the key distribution problem, and modelling the obligations and duties of the parties involved. The cryptographic protocols for key distribution and enciphering/deciphering of the software may be implemented by conventional cryptosystems or public-key cryptosystems.

To give a notion of a software protection scheme based on cryptography and additional hardware, a simple example is outlined from the customer's point of view. This example is in part taken from Albert and Morse, (see references, section 2).

The hardware manufacturer (or OEM) H stores the serial number of each deciphering processor (see Figure 6) and the deciphering key kH in the memory of the tamper resistant deciphering processor.

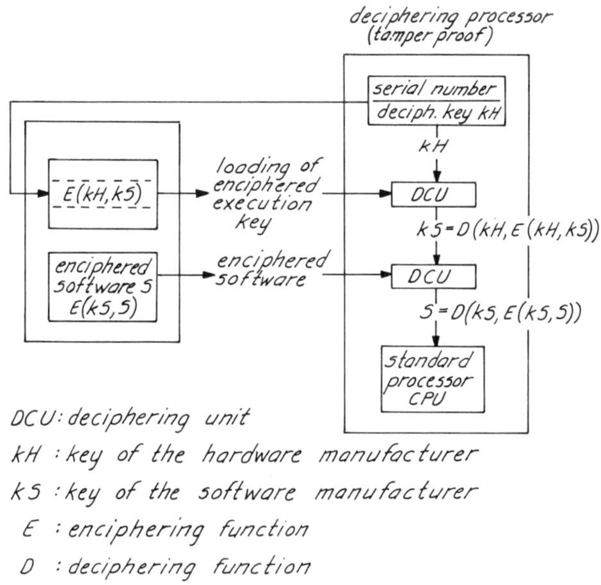

Figure 2.6 Execution of Enciphered Software

Whenever the customer C purchases the software S from the software manufacturer P, C submits the serial number of his hardware deciphering processor. P submits this serial number together with the execution key kS to H for enciphering it with the enciphering key kH associated with the serial number and the decryption key kH.

The software manufacturer P enciphers the software S with the execution key kS associated with S and the program (serial) number. The customer C gets the software that is particularly customized to run on C's processors only, containing the appropriate serial number and deciphering key kH. Figure 2.6 shows how this scheme works at program execution.

There are refinements of that scheme and some variations using chipcards (or smart cards) containing deciphering key kH and/or kS. Examples are given in a paper by Schaumüller and Piller listed in References.

Depending on the implementation, it may be sufficient to encipher only the most important part of the software, e.g. the proprietary algorithm for calculating the net income. Due to the self-authenticating nature of encipherment this software protection method also protects against unauthorized modification, e.g. the software controlling power plants. Moreover, software protection is one remedy against the new computer disease: computer viruses, provided the enciphered software does not already contain a virus.

General key distribution protocols for protecting software using public key and private key cryptosystems are given in a paper by Herzberg and Pinter (see References). Some steps towards a theoretic foundation of software protection are outlined in a paper by Goldreich (see References).

Outlook

The areas described in this section, to which cryptographic techniques can be applied are by no means complete. We mentioned here only two further areas: certified mail, and contract signing. A sophisticated version of digital signatures is used in generating certified mail. One party sends a message to another party. The second party gets the message only if the first one obtains a signed receipt (see Blum in References). Another extention is that the receiver learns absolutely nothing of the contents of the message until the originator gets the signed receipt.

A similar application is contract signing. The cryptographic protocols allow two parties to sign a contract in a secure way similar to the security you get while signing a traditional paper (see Even, Goldreich, Lempel in References).

3 ALGORITHMS

3.1 Introduction

An algorithm is a set of formulas (and rules) describing the way a computation or a task is to be performed. A cryptographic algorithm is an algorithm prescribing the way plain text is transformed into cipher. Since in most circumstances plain and cipher text can be coded as strings of bits, a cryptographic algorithm can be thought of as a mathematical function mapping a string of bits P (the plain text) into another string of bits C (the cipher text) influenced by the secret key K:

$$C=f(P,K).$$

To make deciphering possible, this function has to have an inverse

$$P=f^{-1}(C,K)$$

The number of functions theoretically possible is enormous, in reality limited only by security considerations and by what is possible to realize in soft- or hardware.

In examining and evaluating a ciphering algorithm it is customary to assume that whoever is going to try to break it knows the function f; thus the secrecy is carried solely by the secret key K.

3.2 Theoretical security

It is quite easy to realize that if the size of the secret key, by which we mean the (2-logarithm of the) number of possible keys, is small compared to the amount of data transmitted using one key, then the system can, at least theoretically, be broken using a method known as the Brute Force Attack. In this method the intercepted cipher text is deciphered successively using all possible keys. If the amount of cipher text is large enough, only the correct key will yield a deciphered text with the correct structure (correct parity bits, frame information, readable text...), all others will not.

It is even possible to prove that if the goal is that the attacker gain no information, then it is necessary that the key have the same size as the plain text. Such systems are known as Infinite Key Systems, and have historically been implemented in the form of One Time Pad (OTP) and One Time Tape (OTT) systems.

Unfortunately, Infinite Key Systems are exceedingly impractical in most data transmission systems. Therefore, modern designs accept the threat of the brute force attack and counter it by making the secret key so large that the time needed to execute the attack is prohibitive. This is usually not difficult, since even for a moderate key size (60-90 bits) the time needed may run into millions of years. To decrease the time needed, an attacker might attempt several keys at a time, with the effect, however, that costs will increase (time-cost tradeoff). This possibility will have to be taken into account when designing systems.

Having in this way disposed of the brute force attack, one is left to worry about intelligent attacks.

3.3 Stream ciphers and block ciphers

In order to be able to discuss what can be done to protect against intelligent attacks we shall consider two classes of algorithms, stream ciphers and block ciphers.

Until recently, probably over 95 % of all cipher algorithms in use were stream ciphers, in which a key generator, controlled by the secret key, produces a bit stream (the key stream) which enciphers the plain bit stream by simple MOD 2 addition.

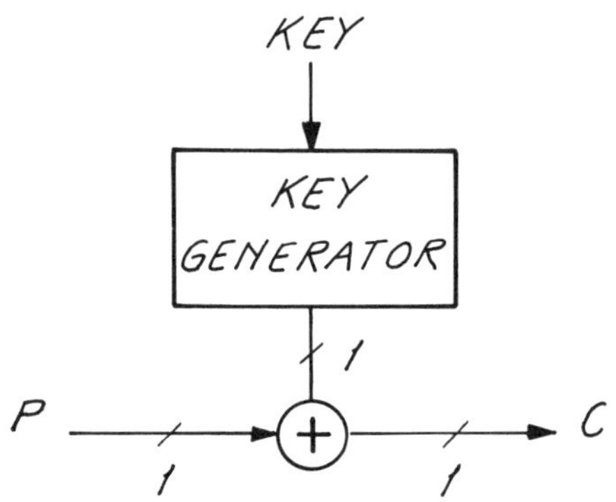

Figure 3.1 Principle of stream ciphers

A stream cipher system thus hides the plain text by changing the bits of it in a randomlike way. An interceptor, who does not know the key, will not know which bits have been changed (corresponding to the occurrence of '1' in the key stream), or which ones remain unchanged ('0' in the key stream). Therefore, to him, at any position of the plain text, a '0' is as likely as a '1', which means that he gains no information about the plain text bit by observing the corresponding ciphertext bit.

An ideal stream cipher would use a physical (true) random number generator as key generator. Since its output cannot be reproduced, however, decipherment would be impossible, unless the whole key stream, with the same length as the plain text, is transported to the legitimate receiver via a safe channel. This procedure, which is an example of the infinite key systems mentioned above, is often impractical, wherefore mostly so-called <u>pseudo-random number generators</u>, with special properties, controlled by a relatively short key, have to be used instead as key generators.

One of the fundamental cryptographical problems can be said to consist in making the key generator produce a key stream which is unpredictable from the point of view of the cryptanalyst, i e for somebody who does not know the secret key. The cryptanalyst, or the attacker, is assumed to have available a certain amount of cipher text, as well as some information about the corresponding plain text. The pairing of these two pieces of information leads to some knowledge about the key stream. That the key stream is unpredictable is taken to mean that it should not be possible, with reasonable cost or in reasonable time, to find out more about the key stream; in particular, it should not be possible to exactly determine a section of the key stream, if partial knowledge of this section is available, nor should it be possible to find the continuation of the key stream, if a section of it becomes completely known.

Another class of algorithms is constituted by the block algorithms. Unlike the stream ciphers, where only one bit at a time is ciphered, in this case whole blocks of bits are treated simultaneously.

In this case the plain information is hidden by the fact that, depending on the key, a cipher text block can be deciphered to any combination of m plain text bits or, if the number of keys is less than 2^m, to as many combinations as there are keys.

If the keys are chosen with equal probability, then to the interceptor observing a cipher text block, all the possible plain text blocks are equally likely to have occurred.

An example of a block algorithm, with block size 64, is the Data Encryption Standard, DES, standardized by the National Bureau of Standards in the United States (see also the chapter on standards).

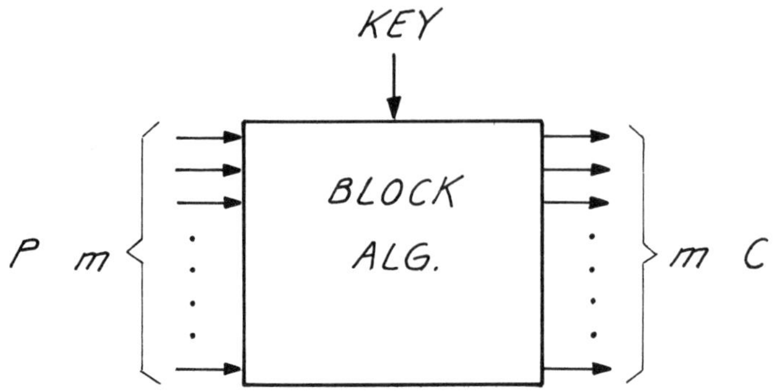

Figure 3.2 Block cipher system

If a message is longer than the block size then it is divided up into a number of sections, each of block size length. These sections are then enciphered and transmitted serially.

Again, the fundamental cryptographic problem is one of prediction: if something is known about the transformation, can further information on it be obtained? For example, given that a certain plain block P_1 ciphers to C_1, can one predict with some confidence how another plain block P_2 will be ciphered, or how another cipher block C_2 will be deciphered, or, given a cipher block C_1 and some information on the corresponding plain block P_1, can more information on P_1 be obtained?

Of course, if possible, the best way to make predictions would be to compute the secret key. Therefore, cryptographic algorithms must be designed so as to make it impossible to calculate, in reasonable time, or with reasonable chance of success, which secret key has been used, if a certain amount of plain text with corresponding cipher is given (known plain text attack).

So far we have assumed that the purpose of the algorithm is to keep secrets. There are other things that algorithms can, and should, do, such as guaranteeing the authenticity of the sender and the integrity of the message: the recipient of the message should be confident that the originator of it is who he claims he is, or at least that he is allowed to send messages, and also that he should be sure that the message has not been tampered with.

The authenticity of the sender is, at least up to a point, also ensured by the non-predictability of the ciphering process; without the key it is impossible to generate a meaningful message. The integrity of the message is not necessarily guaranteed by the unpredictability, however. In fact both stream and block ciphers in their simple forms leave something to be desired on this score. This problem will be discussed more below.

3.4 Examples of attacks

a) Attacks on the secrecy

To illustrate the prediction problem mentioned above we shall consider some easy examples of stream cipher systems.

A counter is a device the states of which are integers less than some maximal value m. The new state is the old one plus one, unless m is reached, in which case the counter is reset to zero. At the beginning of a transmission the counter is initialized to a value determined by the secret key. If we denote the state of the counter at step n by s_n we have:

$$s_0 = \text{key dependent}$$

$$s_{n+1} = (s_n + 1) \text{ MOD } m$$

The counter as such is of course predictable once one state is determined. Thus, if we use for key stream the binary representation of the successive integers, we immediately realize that with knowledge of enough bits of the key stream, the state can be computed. Therefore, we try instead to use only one bit from each state, say the least significant bit, or the sixth order bit. In this case it is mostly not possible to determine the state of the key generator after observing a section of the key stream.

On the other hand, it is not necessary to determine the state of the generator to make predictions; if the least significant bit is used for the key stream it is sufficient to observe one bit to determine the following ones with probability equal to or close to one. In fact, in this case the key stream has a pattern ...01010101..., which is possibly broken only when the counter is reset. On the other hand, if we use the third order bits, then the key stream will consist of alternate sections of 8 zeros and 8 ones. Hence to make the prediction we need only observe one change-over from one section to another.

s_0 0101110
s_1 0101111

s_2	0110000
s_3	0110001
s_4	0110010
s_5	0110011
s_6	0110100
s_7	0110101
s_8	0110110
s_9	0110111
s_{10}	0111000
s_{11}	0111001
s_{12}	0111010
s_{13}	0111011
s_{14}	0111100
s_{15}	0111101
s_{16}	0111110
s_{17}	0111111
s_{18}	1000000
s_{19}	1000001
Order bit:	6543210
	MSB LSB

Figure 3.3 The states of a counter in binary representation

The probability that this event occurs decreases with the order to the bits used. On the other hand, with increasing order one can with ever higher probability predict the continuation of the key stream by guessing that it stays constant.

For this and many other reasons, a simple counter is ill suited as key generator. Instead we try an irregularly stepping counter known as LCA, Linear Congruential Algorithm. Here the step from one state to the next is made according to the formula

$$s_{n+1} = (a \times s_n + b) \text{ MOD } m.$$

If a, b, and m are chosen judiciously the states will not repeat until all possibilities, i.e. all integers<m have occurred.

Much greater sophistication is needed to see that also in this case, if we use bits of a certain order as key stream, it is possible to predict the state of the generator if a sufficiently long part of the key streams is known.

To raise the cryptographic level of the LCA one might consider using for key bits a more general function of the state, not just one of the bits of the state. This would seek to prevent the attacker from ever obtaining 'hard' information on the state of the counter. As an example we could use the function x Æ y = f(x) = a^x MOD p, with appropriately chosen parameters a and p, and use a certain order bit of $f(s_n)$ as key stream bit.

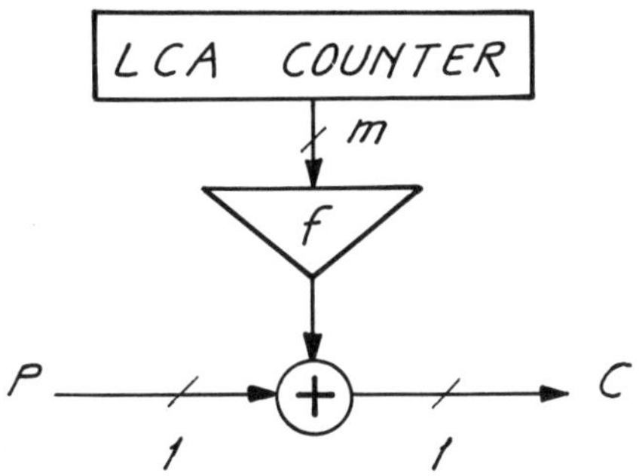

Figure 3.4 Stream cipher with an LCA counter

In principle we could choose to use more than one, or even all, of the bits of $f(s_n)$ for our key stream.

The reason we choose the special function $f(x) = a^x$ MOD p is the following. Suppose all the bits of $f(s_n)$ are used to generate the key stream. Then, if the interceptor were to know a section of the plain text, he could compute the corresponding key stream bits, and hence maybe $f(s_n)$ for some n. If he now could invert the function f, i.e. compute the number s_n from $f(s_n)$, then he would have broken the system, in the sense that he would be able to successively compute s_{n+1}, s_{n+2} ... and thereby also $f(s_{n+1})$, $f(s_{n+2})$,..., which in turn would allow him to decipher the rest of the message. Therefore, if the proposed algorithm is to be considered secure, it must be impossible to invert the function f(x).

Fortunately, it is generally acknowledged that the computational problem: given integers a, p and y, find an integer x such that

$$y = a^x \text{ MOD } p$$

is 'computationally impossible' if the integers involved are large. By this we mean that there are no algorithms which allow one to solve the problem on a computer in 'reasonable' time or with 'reasonable' effort. Of course, the brute force method, i.e. trying all intergers x, is theoretically possible, but would take an 'unreasonable' time.

Going back to the case where we use only one bit of $f(s_n)$, one is tempted to conclude that the security level would then be at least as high as when we use all bits.

Note that for this ciphering algorithm to be practical, the computation of f(x) from x must be easy. With $f(x) = a^x$ MOD p, this requires a number of operations which is proportional to log p, making the method usable in many applications with integers of several hundred digits.

A function f for which it is computationally easy to compute f(x), while it is difficult or impossible to compute $f^{-1}(x)$, is known as a computational one-way function.

The function $f(x) = a^x$ MOD p, thus, is believed to be an example of a computational one-way function.

More advanced stream cipher designs use functions f with memory, so that $y_n = f(x_n, x_{n-1},\ldots,x_0)$, and/or functions f depending on the secret key.

With this example another principle in the design of cryptographic algorithms is illustrated: one tries to make the breaking of the algorithm more or less equivalent to the solving of a mathematical problem of better known level of difficulty. This approach is not without its pitfalls. The equation stated above is thought to be impossible to handle in general. However, for certain a, or certain p, or for certain combinations a and p, the equation may be solvable in reasonable time. Thus such parameters must be avoided, if the algorithm is to be secure.

b) Attacks on the authenticity and integrity of messages

As mentioned earlier, the authenticity and integrity of ciphered messages are in principle ensured by the secrecy of the secret key: without access to it a meaningful message can neither be generated nor arbitrarily changed.

Under certain circumstances, however, this property is lost. This happens when message secrecy is wholly or partially lost, regardless in which way. Stream systems are particularly vulnerable in this respect; if the exact plain text of a cipher message becomes known to the attacker (through theft, espionage, bribery) he can change the cipher text to make it correspond to any other plain text of the same length.

Messages ciphered with simple block algorithms can be corrupted through the changing of the order of blocks, or by substituting blocks from another (known) message.

These threats can be countered by randomization methods, such as including a certain number of random bits in the message, and a checksum, or by randomizing in time, e.g. by ciphering random information continually, and introducing plain text only at random intervals.

Both block and stream algorithms can moreover be protected by chaining procedures which will be discussed below.

3.5 Modes and comparisons

To remedy some of the disadvantages of stream and cipher systems as described above, they can be used in more involved modes.

a) Modes of stream ciphers

Stream ciphers of the type described above can be run in two kinds of cipher feedback modes.

In the simplest type, the counter is dispensed with and the last m bits of the cipher stream are used to address the function f.

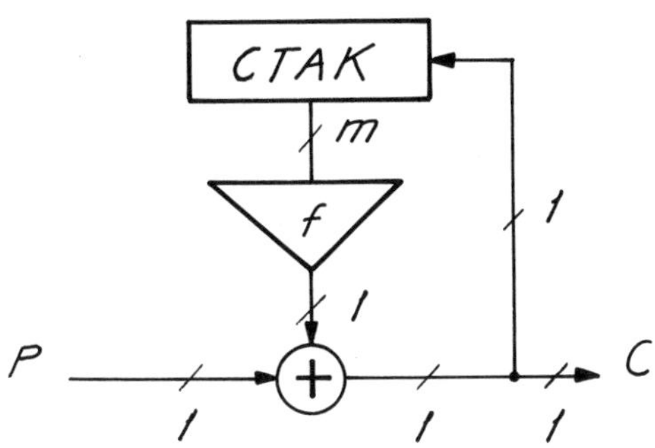

Figure 3.5 Simple feedback for a stream cipher

Attempts to change the cipher text will result in the destruction of about m bits of the plain text and will be detected.

A more refined method, often called cipher autokey, still uses a counter but the output of it is XORed to the feed back bits before being used.

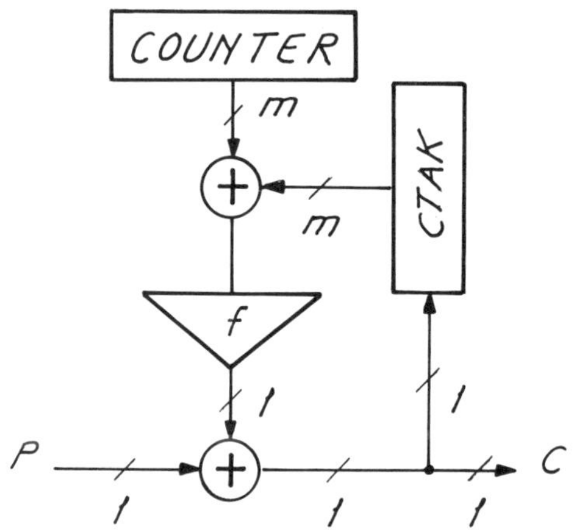

Figure 3.6 Cipher autokey

b) Modes of block ciphers

Block ciphers can be used in several modes. The simplest of these is Electronic Code Book Mode (ECB), in which each plain block is enciphered in turn.

A disadvantage of this mode is that a plain block, if occurring more than once, will always be ciphered to the same cipher block. An attacker, having by some means identified the plain block, can discover when it appears again, or can use the corresponding cipher block and insert it into other messages.

To counter this threat several <u>chaining modes</u> can be used, the most important of which are Cipher Feedback (CFB) and Cipher Block Chaining (CBC).

In the cipher block chaining mode, the plain block is XORed with the previous cipher block before encipherment takes place.

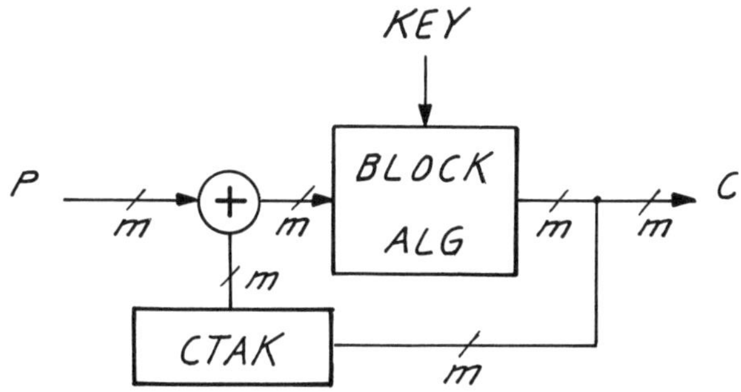

Figure 3.7 Cipher Block Chaining

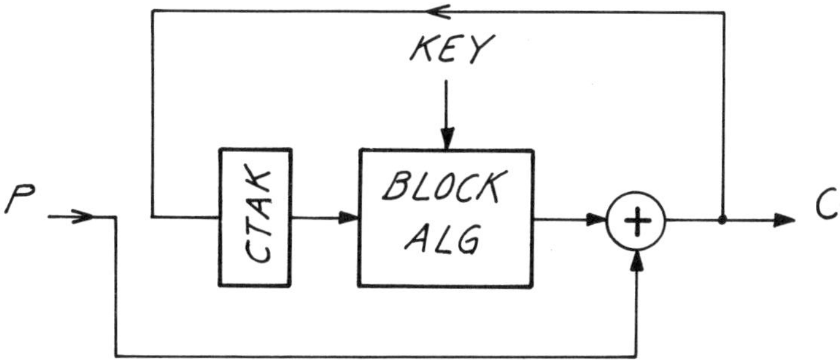

Figure 3.8 Cipher Feedback

Thus the cipher block corresponding to a certain plain block is also a function of all the previous plain blocks. The same is true for the cipher feedback mode, where ciphering is performed by XORing the plain block with the output of the block algorithm, the input of which is the previous cipher block.

The feedback need not consist of a whole block. In the extreme case only one bit is fed back, and only one bit of the output is used:

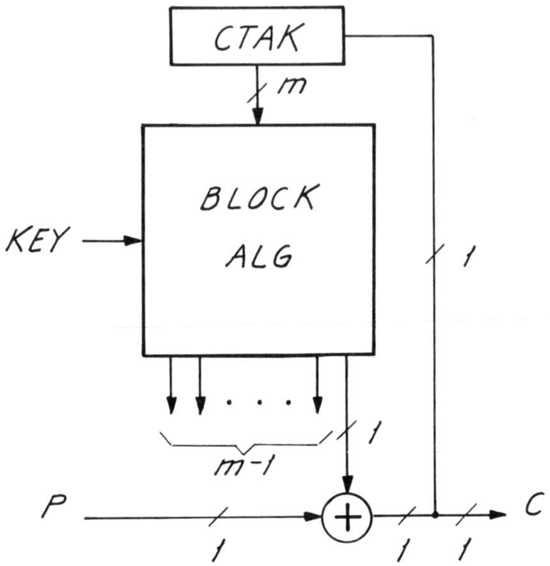

Figure 3.9 One bit Feedback

The analogy with a stream cipher with cipher text feedback is obvious.

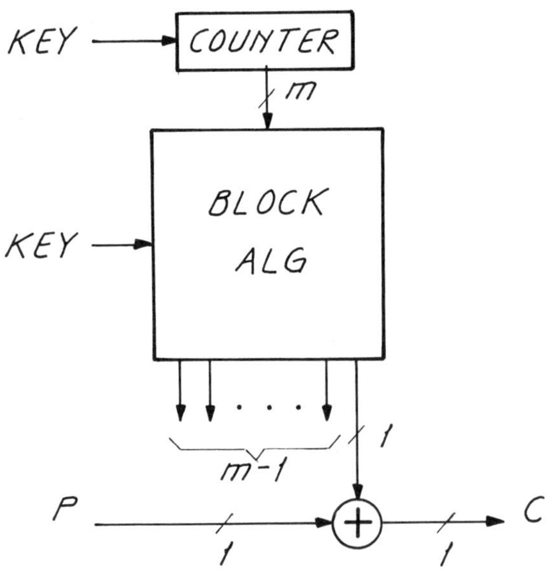

Figure 3.10 Stream cipher using a block algorithm

A block cipher system can even be turned into a stream system by letting the block system act as the function f in section 3.4 a. See Figure 3.10.

c) Block/stream comparison

The mode examples above show that the differences between block and stream cipher systems are more of a practical than of a theoretical nature.

The essential difference lies in the fact that in the ECB, CFB, and CBC modes one deciphered bit depends on, i.e. is a function of, many cipher blocks. This has important implications, in particular for error propagation. Thus a bit error in the cipher will cause one or two plain blocks to be destroyed, something which in data transmission environments is considered an advantage: a single bit error is unlikely to go undetected. In different circumstances it is considered a disadvantage: a singel bit error can ruin an otherwise perfectly readable message.

Cryptographically, block algorithms are more general and hence can in principle be made stronger. On the other hand, stream ciphers are widely held to be easier to analyse and evaluate, and are usually felt to be sufficiently secure, provided the authenticity and integrity problems can be handled by other means.

3.6 Crypto variables

a) The secret key, weak keys

In the discussion of the brute force attack we tacitly assumed that all possible keys had the same probability of being used. If this condition is not fulfilled the security will decrease, as will be readily understood: the attacker will try the more likely keys first, thereby making the average search time shorter.

Thus the general rule is to generate keys randomly with all possible keys having the same probability.

Since there are no rules without exceptions, we hasten to note that for some algorithms there might be 'classes of weak keys', which for one reason or another should be avoided. In practice such classes are important only if they are large; the fear of single so-called weak keys is more psychological than real.

b) Randomizing vectors

Up to now we have considered the secret key, i.e. the parameter controlling the algorithm, to be one entity. In many systems, however, it consists of several parts, some of which may not even be secret.

To illustrate, we consider first stream ciphers. To use the same controlling parameters for two different messages would mean enciphering with the same key stream twice, and this could lead to loss of security. On the other hand we would like to use the same secret key for many messages. To overcome this dilemma we split the control parameters into two parts. One part is the secret key, the second one is known as message key, randomization key, or initialization vector. The second part can be randomly generated by the equipment and transmitted. Before being used in the algorithm it is often mixed with, or 'enciphered by', the secret key, in order that the relationship between two key streams generated by the same secret key and different but known message keys be not exploitable.

In cipher systems with chaining, a similar effect can be obtained by e.g. letting the message key be the first block of the message (plain or cipher), thereby randomizing the encipherment of the subsequent plain blocks.

3.7 Public key systems

By an asymmetric cipher system we understand a system in which the keys used for enciphering and deciphering are different:

$$C = f(P,K_e)$$

$$P = g(C,K_d)$$

Evidently there must exist a relation between K_e and K_d:

$$K_e = f\,(K_d)$$

$$K_d = f\,(K_e)$$

If f is a computational one-way function, such a system can be used as a <u>public key system</u>: K_e can be made public, and all and sundry may encipher message using it. Because of the one-wayness of f, however, only the one having access to the secret key K_d can decipher these messages. Of course, we keep our usual security requirement concerning the function f.

There is essentially only one public key system believed to be secure, namely the RSA system. In this, the public key consists of integers K_e and m and enciphering is performed according to the formula

$$C = P^{K_e} \text{ MOD } m.$$

In order for this to make sense one has to translate the plain text into a sequence of integers<m, e.g. by interpreting sections of the plain text (in bit-string form) as the binary representation of integers.

The deciphering formula is:

$$P = C^{K_d} \text{ MOD } m.$$

Using number-theoretic arguments one can show that it is easy to construct K_d from K_e if the prime factors of the integer m are known. If these factors are not known the construction is computationally difficult, at least according to the current state of knowledge. Hence the security of the RSA system depends on how difficult it is to find the prime factorization of integers.

At the time of writing, the state of this art has reached 84-digit numbers; these can be factorized in a matter of hours on a large computer. For this reason, it is recommended that the RSA system be used with integers of several hundred digits. To avoid that special factorization mehtods be applicable, rules which restrict the choice of m, and which are best left to the specialist to ponder, should be adhered to.

Because of the public nature of the RSA system, one can attack it without intercepting any messages, in fact none need ever have been sent!

The RSA system is another example of how the breaking of a crypto system can be made (more or less) equivalent to the solving of a mathematical problem of simple structure.

3.8 Evaluation of cryptographic algorithms

The evaluation of cryptographic algorithms must be left to the specialists. However, the general user often has several options when proceeding to choose algorithms.

a) Choose a published algorithm, like DES. Philosophy: a public known algorithm will have been scrutinized by a large number of specialists, many of whom can be expected to have published their findings.

b) Trust the manufacturer. Philosophy: A well-known manufacturer has long experience and is unlikely to risk his reputation by selling insecure equipment.

c) Trust a private consultant. Philosophy: A trustworthy consultant with experience can be expected to make a reliable evaluation of algorithms offered on the market. To official or semiofficial users, a special type of consultant is sometimes available. In many countries government cipher bureaus will give advice on algorithms, at least to public or semi-public organisations.

d) Trust the government. In some countries, the government will prescribe which algorithms to use, at least in government-related circumstances (USA). The user is given no information on the algorithm. In others, government permission is required for private organizations to use cryptographic equipment (France). In these cases permission will not be given unless the algorithm is already known to the government, or is presented for examination. The user is sometimes allowed to receive information about the algorithm he is using.

4 PROTECTING THE PROTECTION

4.1 Key Management

Introduction

Data encipherment is now well established as an essential feature of civil computing and communication systems operated by such organisations as banks; it is essential because it is relied upon for provision of privacy and integrity of data, both of which are absolutely vital for the maintenance of confidence in these systems, without which modern business could not be conducted.

Use of encipherment is therefore very widespread and, in the main, relies in this particular context upon published algorithms, several of which are the subject of standards activities. Banks have many times made it plain that they are in favour of using algorithms in the public domain. Two classes of algorithm are involved, those of the symmetric type with the same key required by the encipherer and decipherer of data, and those of the public key type, where different keys are required for the two functions. Because the encipherment algorithms are not secret, the whole of the security structure depends either on the secrecy of the encipherment keys (in the case of symmetric algorithms), or on the integrity of the public keys and secrecy of the private keys (in the case of public key encipherment algorithms). Secrecy of symmetric encipherment keys and integrity/ secrecy of public/private keys is thus central to the survival of the systems to which we refer. For this reason the management of encipherment keys of all types is a subject of the highest importance to which much attention has been given and which has seen some important developments in recent times. The purpose of this section (derived largely from National Physical Laboratory report DITC 70/86) is to consider the safe ways in which such keys can be managed.

We shall be considering methods of key management for applications using the Data Encryption Standard (DES) algorithm, as an example of a symmetric encipherment system, and the RSA algorithm, as an example of a public key cryptosystem. In some situations it may be found useful to employ a combination of the two types of algorithms. Much of what we have to say can be generalised to include key management for other symmetric and public key algorithms respectively.

Key management, in our definition, refers to methods of selecting keys and to ensuring that they are available at the right times at the points where they are needed, and not available in any other circumstances or to any unauthorised parties. It is also relevant to consider ways in which keys may be destroyed when they are no longer current or if some breach of security is detected. The following sections therefore refer to selection, safe transport, protection at the point of application and destruction of keys. In addition we shall address the various roles that can be played by central key servers and also the special needs for key management in Open Systems Interconnection.

Because of the breadth of the subject we are addressing we do not attempt in this to give a fully detailed description of any particular feature or method of key management. Reference should be made to source documents (cited in this text) when this level of detail is required. Surveys at an intermediate level of detail can be found in Davies and Price (see section1 in References) and Price (see section 3 in References).

Selection of Encipherment Keys

Keys may be selected by the immediate user or they may be obtained from some third party source. Selection by the user implies a strong degree of control exercisable by the user, but also the need for the user to possess the means for safe key generation, which may, particularly in the case of public key sets, be non-trivial. Selection by a third party implies, of course, the safe transport of keys from the point of generation to the point of application, a subject which we shall explore in the next section; it also implies trust in the integrity of the third party.

Selection of DES keys

DES keys must be unpredictable by a potential enemy and must therefore be chosen by some reliably random or pseudo-random process, either a true noise source or a suitably strong pseudo-random number generator. Even manual coin tossing has been known to be used for this purpose, generating the keys bit by bit, but an automatic process is bound to be much more convenient.

Before a chosen DES key is used, it must be tested for weakness. It is well known that the DES algorithm has four weak keys, whose effect is that re-encipherment of a ciphertext with a weak key restores the plaintext. There are also six pairs of semi-weak keys, whose effect is that encipherment with one key of a pair followed by encipherment with the other key of the pair restores the plaintext. Just to try re-encipherment of ciphertext could be a very inexpensive step for an intruder to attempt, possibly leading to disclosure of plaintext if a weak key has been used. It is therefore advisable that tests for accidental selection of one of these keys should be made. Their identity is given in a number of texts, including Davies and Price.

It can be argued, of course, that a potential enemy may have no idea that a weak or semi-weak key is being used, and that therefore he gains no advantage from their accidental use. However, in the case of the weak key, captured equipment with encipherment key still installed could be made to re-process ciphertext, thus achieving the desired second encipherment that yields the plaintext. It is generally recognised that weak and semi-weak keys should be avoided in practical key selection.

It is known that other DES keys have certain particular properties leading to regularities in the internal working of the algorithm; see, for example, Kolata (see section 3 in References). However, no cryptanalytic advantage has been identified in a convincing manner for these other 'special' keys, and it is not normally considered necessary to avoid them in use.

Selection of RSA keys

The first consideration in selecting an RSA key set is that of parameter size. The pair of primes, p and q, whose product is the modulus, m, used for the modular exponentiation which is the embodiment of the RSA algorithm, must be chosen large enough for factorisation of their published product to be infeasible to the potential enemy. If p and q are discovered, then security of a system using them is totally lost. Size of p and q is therefore related to the feasibility of factorising large composite numbers. This is a subject which has received attention from mathematicians over the last few centuries, especially since the time of Gauss, and has provoked intense interest since the publication of the RSA algorithm.

The present limit of attainment (June 1987) is factoring RSA moduli of 88 digits; the method, based on a quadratic sieve algorithm, is implemented on a group of Sun workstations and requires one month of dedicated running.

It is possible that one hundred digit factorisation may be feasible by the end of the decade if present predictions of progress in hardware and algorithms prove correct.

Therefore it is often suggested that the RSA modulus should be chosen to be well in excess of eighty digits and the original RSA paper was written in terms of 512 bit moduli (about 155 digits). For many applications this may be considered excessively large, on the grounds that availability of a large machine capable of factorisation at this level will be extremely limited, in practice leading to adoption of smaller moduli; smaller modulus implies a lesser processing load in carrying out the algorithm, which is a potential advantage with this demanding computation. On the other hand, for applications requiring very high security, then a modulus size even larger than 512 bits may be required. If we assume that a 512 bit modulus is to be used, then we need to seek candidate values of primes p and q in the region of 77 digits.

Size is not the only consideration in selecting these prime numbers. In order to avoid such weaknesses as 'short cycling' of the RSA algorithm, it is necessary to lay down a set of rules with which selected primes must comply. These have been clearly expressed in a paper by Gordon (see section 4 in References), where Gordon shows how to derive the large primes from smaller primes, obeying the selection rules and obtaining results with economy of computation.

As we have indicated earlier, the choice between direct user selection of RSA parameters and obtaining these from some specially provided service centre is one that must be faced. Use of a centre implies that this has to be trusted by the users not to betray the selected values and not to issue the same key set to more than one user. It also implies that the secret parameters must be transported securely to the user location.

Because the RSA algorithm is demanding in computational terms, various suggestions have been made from time to time with the object of lessening the processing load in particular implementations. One such suggestion was to use an exponent of 3 for the public key, with an arbitrary exponent, determined by the usual rules, for the secret key. The advantage of this is that encipherment is then a very cheap operation, being the taking of a modular cube. Decipherment processing is not reduced. Provided that this limitation on choice of public key is used sensibly, the suggestion has merit. There is at least one pitfall that must be pointed out. This arises if the same plaintext is sent to different destinations enciphered with exponent 3 and different moduli. Then the plaintext can be found with very little computational effort, though the secret key parameters are not thus compromised.

Another suggestion relating to choice of RSA key sets has been the use of a common modulus for a group of users, with independent public and secret exponent pairs for each individual user. Because each user will need to know his own secret key (and, of course, knows his own public key), it is possible for users, knowing the public keys of other users, to derive the secret keys of the latter with little computational effort. Therefore the suggestion to use common modulus operation within a user group needs careful consideration. If the group is one where mutual trust exists, then there seems to be no point in using distinct key sets for the different users. If the group is not one of mutual trust, then use of the same modulus is dangerous.

Transport of Encipherment Keys

As we have indicated in the introduction, secret encipherment keys need secure transport to the points at which they are going to be used. Perhaps the most obvious way of transporting a secret key is to use the channel which is to be protected by the key. Because it must be assumed that the channel may be tapped, it is essential to protect the key in transit by enciphering it. This demands that there shall be an operative key in place before the transfer takes place and provision must be made for installation of this key. For secret keys, therefore, the highest key in a hierarchy cannot be sent over the communication channel itself and requires a different system to handle it. Generally speaking there is no alternative but to arrange key input to encipherment devices by human intervention using a different means of transport. For a public key system, the problem is quite different; here it is necessary to make public keys available where they are needed and to ensure their integrity; by definition secrecy is not required for public keys.

Physical Transport of Secret Keys

Safe movement of a key from point of generation to point of use, where the key is at the highest level of a secret key hierarchy, requires some kind of safe physical transport. At the most primitive, the key may be printed on paper, sealed into a packet and sent through the ordinary mail; at the destination the packet is opened and the key input to the encipherment unit by human operation of a keyboard. Such a system is, of course, open to various kinds of abuse and may also be error prone; it is, however, the simplest way in which to handle secret keys of this type and is acceptable in many circumstances. Errors in key installation can easily be detected by means of agreed test transmissions of enciphered text.

In order to increase the security, the key can be split into two or more parts, which are sent separately. Thus the security is not compromised if one of these parts is intercepted, or one of the legitimate receivers of these parts has been bribed into disclosing it. The security can also be enchanced by hiring trusted persons to transport the keys (couriers).

For systems requring higher security it is essential to use special hardware. This may take the form of a 'key gun' which is designed to accept keys from the generation equipment and to issue them to encipherment units as required, the gun being transported by courier or, possibly, by mail. It is vital that the contents of the gun shall not be available to any unauthorised parties. Therefore a good design should ensure that the gun can recognise the unit into which it is plugged. Keys should only be accepted from a recognised and authorised key source and only issued to recognised and authorised encipherment units. If this is correctly realised, then there is no reason why a key gun should not contain a number of keys to be installed successively in different units. It may be thought necessary to delete each key from the gun as it is issued to the appropriate encipherment unit. Physical security of the key gun is very important; any attempt to break into the unit should be detected and the response of the unit should be to delete all secret information it contains. Appearance of the unit should be very distinctive and difficult to imitate. This will help operators of the key generation source to recognise genuine units and to reject any impostors. A possible method of making the appearance difficult to imitate is to use a complex, but recognisable, hologram securely bonded to the unit.

Clearly, an interface must be defined for the key gun and encipherment unit; the interface is most likely to be electrical, but some manufacturers use optical transmission. One such electrical interface is described in the US federal standard for physical security of encipherment modules (Federal Standard 1027).

In conclusion of this sub-section, we comment that many key management schemes for secret keys assume that the stage of highest level key installation has already taken place. Such schemes do not always address the problem of initialisation of the secure system.

Secure Key Transport Using the Protected Channel

Once high level keys are installed at the two ends of a communication channel, it is possible to exchange in a secure manner keys which will be used to protect data flowing on the channel. It is customary to talk of 'data encipherment keys' and 'key encipherment keys'; other terms for these parameters are respectively 'session keys' and 'master keys' (see Figure 4.1 and Figure 4.2). Many commercial encipherment units make use of only two levels of key hierarchy; a pair of units contains a master key which is used to exchange session keys. In its simplest realisation this scenario assumes that one unit of a pair is capable of generating session keys for transmission to its partner unit; where this is not possible, the key must come from some other source. Some units are capable of operating in multi-drop mode, with a master unit maintaining individual session keys for communication with a number of dependent slave units.

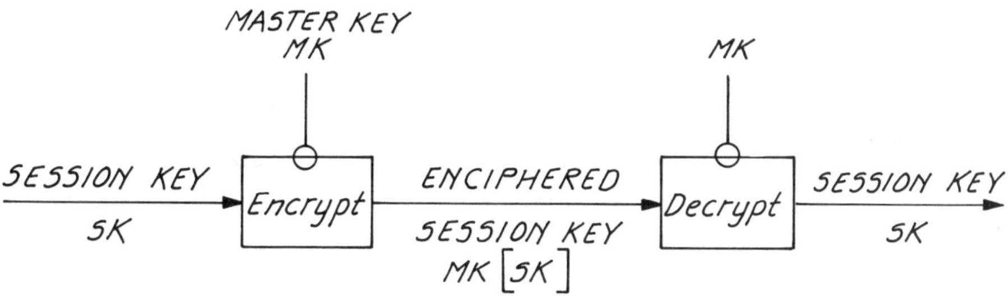

Figure 4.1 Distribution of session key under master key

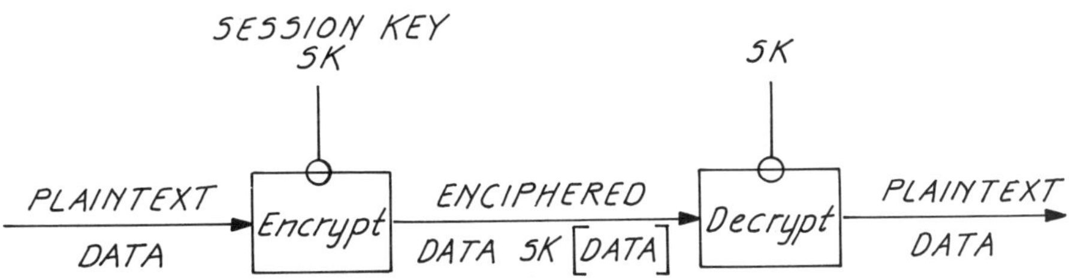

Figure 4.2 Encipherment of data under session key

Not all secure communication systems using the principle of session and master keys are as conceptually simple as that we have just outlined. A notable example of a complex sytem is that of IBM (see Ehrsam, Matyas, Meyer), designed for use in conjunction with its Systems Network Architecture communications hierarchy. This system has three main aspects, secure communication between a host computer and a set of satellite terminals, secure file storage at a host and secure communication between peer hosts; it does not seem well suited for direct user to user secure communication. It is not our aim to give a full description of this system in this section, but to consider its salient points.

In the IBM key management system each host has a cryptographic facility, which is located in a protected storage and processing area. Each terminal has a simple cryptographic module, containing at initialisation a terminal master key (KMT). This terminal master key is also held at the host, under encipherment with a higher level master key (KM1). Session keys (KS) are generated with a pseudo-random process in an unprotected host location; each pseudo-random number is treated as though it were a session key enciphered under a master key (KMO) and, when required for use, passed in this form into the cryptographic facility, where it is subjected to key translation from encipherment with master key KMO to encipherment with the terminal master key KMT. In the latter form it is available for safe transmission to the relevant terminal, where it can be deciphered with KMT and installed for use in protecting data on the host-terminal link.

Secure file storage is achieved by a very similar system in which file keys take the place of session keys and for which there is a different hierarchy of master keys. File keys are protected under a file master key, which in turn is held safe under master key KM2. It should be noted that KM1 and KM2 are related to KMO by a regular system of bit inversion; this does not impose any known additional weakness on the security of the system. In any tests for weak keys in the IBM context, key variants are taken into account in order to avoid accidental choice of a weak key by bit inversion. KMO, KM1 and KM2 do not appear in plaintext outside the host cryptographic facility.

Communication between peer hosts requires the existence of a transfer key which allows translation between parameters enciphered under the master keys of one host to encipherment under the master keys of the other. In the IBM system there is a close analogy between this transfer key and the terminal master key mentioned above.

Access to the cryptographic facility in each host must be very carefully controlled to prevent illegal operations on the part of system users.

Another variation on the theme of session and master key schemes is provided by the National Bureau of Standards with their key notarisation technique (see Smid, References section 3). This scheme is centered on network nodes with special secure peripherals, known as key notarisation facilities (KNF), serving local satellite terminals. It is designed to provide the following:

1) secure communication between any two terminal users (on-line)

2) secure communication via enciphered mail (off-line)
3) protection of personal (non-shared) files
4) provision of a digital signature capability

For inter-user file transfer the NBS scheme uses session keys, generated within the KNFs, which are conveyed to the distant sites under master key encipherment in which the end-user identities are incorporated by exclusive-OR function in the master key (unique to the host pair), thus yielding a variant of the master key, in order to provide evidence that the session key has been used between a particular source and destination user. It is this feature of user-related keys that embodies the 'key notarisation' principle which is characteristic of this method of key management. Because the encipherment key is effectively polarised, it is possible with this method to prove that enciphered data has flowed from a particular source to a particular destination; this therefore provides a kind of digital signature and, it can be argued, a weak kind of confirmation of receipt (at least the enciphered data can only be meaningfully delivered to a defined receiver).

The NBS scheme demands special hardware at every node and does not cater for security between node and dependent terminal, which must be provided by some other means, outside the notarisation scheme, if full security is to be achieved.

The role of central key servers

In the preceding sections the assumption has been made of the existence of master keys between users who wish to communicate securely. In a small community of users this is a reasonable assumption, with only a small number of distinct master keys needing to be held by each individual user. However, where the community of potential communicators is large, as, say, in a world wide network context, then the burden of maintaining master keys for each possible communicating pair becomes very onerous. In these circumstances a different solution must be sought. Creation of a central key server or a set of such servers can make matters much easier.

The simplest role for a central key server is that of key distribution centre for symmetric encipherment keys, where the server has the responsibility of generating session keys on demand by users (see Figure 4.3). In order to transport these keys securely to the calling and called users, master keys must be maintained between the central server and each individual user; the number of such master keys is, of course, much less than the number required between all user pairs. Provision must be made within the protocol for preventing session key substitution or replay; this requirement is usually met by maintaing a system of key numbering and/or time/date stamping. It is also usual for the latter stages of the protocol to include confirmation that the right session key is installed by the two communicating parties.

A slightly more complicated role for the central key server is that of key distribution centre for symmetric encipherment keys, but using public key encipherment for the actual distribution function.

Here it is not necessary to establish master keys between the centre and all users.

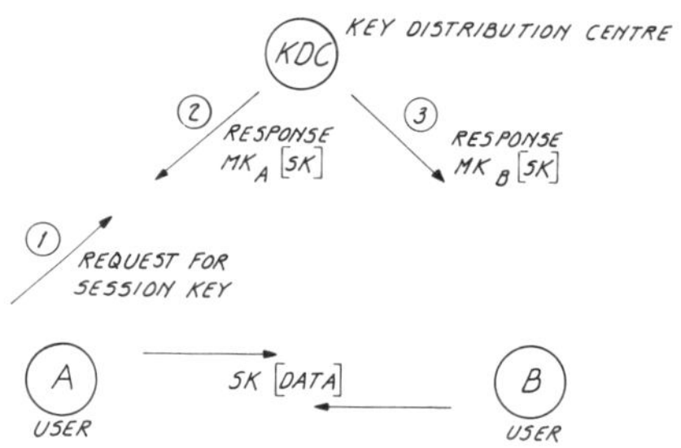

Figure 4.3 Distribution of session key from key distribution centre

That which must be established is universal and certain knowledge of the public key of the centre; it is necessary also for the centre to know the public keys of all participating users, which will have been communicated to the centre by some trusted means. (Trusted, but open, communication, as is required in this case, is often easier to ensure than secret communication.) Once this is done, requests can be made to the centre for symmetric session keys; secrecy of these requests can be maintained by encipherment under the centre's public key. The requests are met by choosing the key at the centre, enciphering it with the public key of the respective calling and called users, and signing the result using the secret key of the centre. The keys can then be sent out to the users who will check their integrity with the public key of the centre. This protocol also should be concluded by testing the installation of the right session key with test enciphered transmissions.

The rationale of transmitting symmetric session keys using public key encipherment and signature is based upon the convenience of public key working, despite the heavy cryptographic computation involved, the lesser problem of managing public keys, and the efficiency of the DES symmetric algorithm in encipherment of data as compared to that of a public key system. A hybrid scheme of this kind can make good sense in some circumstances.

It is, of course, possible to operate a secure communication system using only public key cryptography. In this case the centre takes on the role of key register, where reference is made by any user wishing to call another user with an enciphered call. The centre issues public keys on request, signing each reply with its secret key for checking on receipt by the user. Communication between users is then carried out using the public keys available at each.

The route taken by secure key-carrying messages from the centre to the users can be one of several. The most obvious method is for the required keys to be sent to the caller from the centre. The caller then places the call, including in the calling protocol format the enciphered information required at the called user. An alternative is for the centre to send the enciphered keys independently to calling and called users; this may lead to a lack of synchronism in which the keys may arrive at the called user later than the call request, possibly creating confusion. Yet another way of establishing the keys at the required locations is for the response to be sent by the centre to the called user, who then, on receiving the call request, includes the encipherment parameters required by the caller in the call response.

In the foregoing discussion, where the centre is involved in setting up a symmetric encipherment session key between parties, the centre has the role of choosing the key and then making sure that it is made available at the right places. This deprives the calling user of the right to influence the choice of session key.

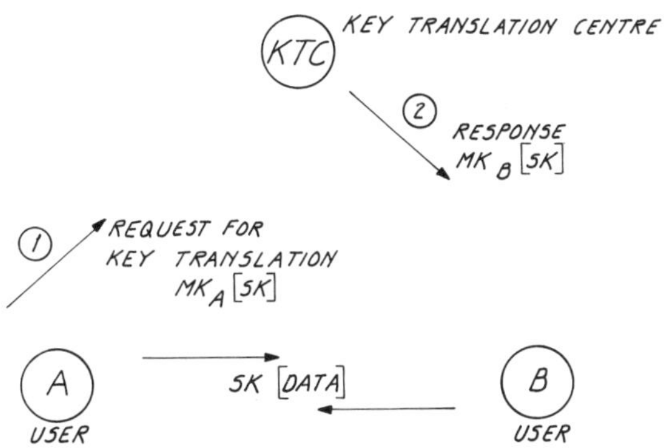

Figure 4.4 Distribution of session key through key translation centre

Where the users of a system require to reserve the right to choose session keys for themselves, it is possible to involve a central key server in a somewhat different role (see Figure 4.4). The problem is still that of wishing to avoid having to create too many master keys between all possible communicating pairs. The role of the centre becomes in this case that of key translation.

Operation of a key translation centre is very similar to that of a key distribution centre. For a system using only symmetric encipherment, the calling user chooses a session key, enciphers this under the master key he shares with the centre and sends it to the centre together with an indication of the called user identity. The centre deciphers the key from the message it receives and re-enciphers it under the master key of the called user. It can be sent to the called user either via the calling user or direct, as desired.

Where a public key system is used for key transmission, but symmetric encipherment for data protection, the session key is again selected by the calling user, sent to the centre enciphered under the centre's public key, with the request that it be sent on to the called user. At the centre it is deciphered and then re-enciphered under the public key of the called user. For maximum integrity the message from caller to centre should be signed with the secret key of the caller and the message from centre to called user should be signed with the secret key of the centre.

Key distribution and key translation centres are likely to be organised to serve limited communities of users. It is therefore quite possible that communicating parties may be located within the domains of different centres. There is then a need for inter-centre keys and an appropriate key-exchange protocol between centres.

Note that the identity of the session key is known at either key distribution centre or key translation centre, whichever system is being used. Therefore the trusted integrity of such centres must be assured.

A fuller discussion of the role of a central key server may be found in the NPL report DNACS 43/81, April-81. It should be noted that the ANSI standard for financial institution key management (wholesale), X9.17-1985, specifies use of key distribution centres, of key translation centres and also of key notarisation. It is becoming increasingly common for European banks to require that encipherment equipment designed for their use should conform to this standard, which is also being used as the base document for an ISO standard in course of preparation.

Key management in Open Systems Interconnection

The discussion up to this point has tacitly assumed that communicating entities and key servers are all at the same logical level in the communications hierarchy, with calls being made between them as required. In practice it is becoming increasingly common to speak of a communication hierarchy of several levels. Within the International Standards Organisation a communications architecture, known as Open Systems Interconnection (OSI), has been defined which separates the network functions into seven distinct layers, ranging from the physical layer to the applications layer.

There is no doubt that data security capability must be provided for the protection of sensitive traffic within OSI. The 'open systems' characteristics of OSI imply openness of access to approved users, but do not imply that data shall not be protected in this context.

The OSI architectural model was standardised in 1984 and since that time work has been proceeding on a second part of the model definition which will be concerned with security. The document arising from this work has by now attained the ISO status known as 'draft international standard', which is an advanced stage in the progression to full published standard. The document addresses many aspects of applying data security techniques in the OSI context; its purpose is to give guidance on how security can be achieved within OSI - it does not seek to lay down implementation details. In particular it is recognised that provision must be made for key management.

The whole OSI philosophy is built around the concept of reliable communication between peer entities, existing at the same levels in the model; for secure data transmission, encipherment, authentication and signature will be applied between communicating peer entities. Information in OSI must also pass up and down the hierarchy across layer boundaries. Where key management is concerned, it seems appropriate to establish a single key management server responsible for keys at all the required levels. There is therefore a need to develop a system of key requests and responses that can flow between communicating entities at various levels and the key server. The latter will probably be located at a high level in the hierarchy. Special communication channels will need to be created for this kind of traffic; in their turn these channels must be afforded appropriate protection. This is an aspect of the security architecture that has yet to be developed to a meaningful degree.

Key management for special applications

In some system designs special considerations must apply to the key management methods. One such case is that of point of sale (POS), where customers will make purchases using tokens, in the first instance magnetic stripe cards. There is the need to check the customer identity, often using a Personal Identification Number (PIN), and creditworthiness, to detect stolen or otherwise blacklisted cards and also the need to transfer funds from the customer's account to that of the retailer.

PIN checking may be carried out at the terminal or it may be referred to the card issuer's central processing system. If the terminal is to be totally off-line, then it must contain PIN validation keys; these are very valuable commodities and require strong protective measures, which it may not be easy to provide within an inexpensive point of sale terminal. The penalty to a system operator of disclosure of a PIN validation key is that of facilitating the manufacture of false cards with valid PINs.

For an on-line POS terminal there are two safe ways of proceeding. If PIN checking is to be at the terminal, the centre may need to supply parameters that allow this to happen. If checking is to be at the centre, then secrecy must be applied to the PIN in transit. In both cases it is essential to set up an encipherment key to protect parts of the procedure; this may be used directly for encipherment or it may allow message authentication to take place. It is generally accepted that the key used should change for each transaction, thus hindering a cryptanalytic attack; indeed, this key is often referred to as 'transaction key'. The transaction key should not allow the reconstruction of the PIN validation key. The latter is not held in the terminal and therefore the requirements for physical protection may be somewhat eased.

Conclusions

It should be clear from this discussion that key management for data security is a multi-facetted complex problem. It is, however, vital that this problem should be adequately addressed in carrying out any secure systems design. Not to do so is to create a possible illusion of data security, whilst, in reality, leaving many gaps through which attackers can penetrate the system.

The problem, though itself complex, does not necessarily demand a complex solution. In many cases it is possible to devise an elegantly simple, yet effective, method of managing encipherment keys. We predict that combinations of symmetric and public key ciphers will be found increasingly in future key management systems.

4.2 Crypto environment

Physical protection

From the preceding parts of this chapter it follows that several objects concerned with encryption must be protected physically against unauthorized access, eg

- encryption keys printed on paper
- "key guns" loaded with encryption keys
- ciphering equipments with installed keys
- keys for physical locks of ciphering equipment
- computer programs for encryption with or without encryption keys loaded

Protection of installed keys

We have considered the safe selection of keys and secure transport to the points at which they are to be used. At the immediate point of use the key must be in cleartext and it is therefore very important to pay proper attention to protection of the key at this point against improper access. This demands that the environment in which the key is used must be equipped with tamper resistance facilities. In the early days of civil use of encipherment it was customary to refer to 'tamper proofing' of equipment, but this is now seen as unrealistic because of the implication of absolute security; tamper resistance implies the property of making improper access too expensive for a practicable attack to be mounted.

The concept of tamper resistance has two aspects - tamper detection and tamper response. Detection is intended to identify a situation in which illegal access is being attempted, whilst response implies the action which the equipment is designed to take in reply to such an attempt. We shall not go into detail in respect of the methods by which these requirements may be met. Suffice it to say that the end result should be that any sensitive parameters held in such equipment should be totally deleted if tamper attempts are detected. ANSI standard X9.17 requires that one thousand overwrite operations with random data be applied to any store of sensitive parameters (following an initial conventional reset operation) in the event of tamper detection.

Destruction of keys

No encipherment keys should be used for an indefinitely long period. A policy must be created for any application which determines the permitted lifetime of all key types. For example, it may be considered that session keys should have a much shorter lifetime than master keys. The rationale for this view is that session keys are used to protect data, which potentially provides an attacker with more material on which to base an attack. Master keys, on the other hand, are used on rare occasions for key exchange, where the keys are unpredictable and provide little or no material of cryptanalytic value for the attacker.

There is therefore a need to replace keys at appropriate times. In circumstances where highest security is demanded it may even be necessary to apply multiple overwrite to the old keys being displaced. Encipherment equipment must be designed to respond to key replacement commands in a trusted manner.

However, since all data, including keys, is easily copied in any computing environment, there is no way in which it can be guaranteed that successful key erasure has taken place in equipment not under the direct control of the entity issuing the command for erasure. The best that can be done is to audit the software and hardware of all encipherment equipment concerned with key use and management, ensuring that the designers intentions are being carried out. Physical protection must be provided in such a way that an audit cannot be falsified.

It is also necessary to remember that keys often exist in other forms before they are transferred to computer media. The necessity of careful, orderly destruction also applies for these other media. Paper, for example, must be thoroughly shredded.

Protection of equipment

Often crypto equipment has to be placed where it can be reached by unauthorized persons. In such circumstances there is an absolute need for the kind of tamper resistant equipment that has been briefly described above. In other cases, however, encryption is carried out with the help of a general computer. Such machines are never tamper resistant.

If encryption keys are stored in or transferred through the memory of a general computer, then great attention must be paid to the security of the computer and its operating system. It must of course be impossible for unauthorized persons to read, alter, or erase the keys or to inject false keys. Corresponding security measures must also be used for encryption software.

In most applications some kind of "handshaking" takes place before any sensitive data are transmitted. In such systems it is pointless to install a false key at one end of a transmission link except as an act of sabotage. If, however, sensitive information might be transmitted before it is clear that the correct key is installed (or if sabotage is a realistic threat) the encryption device must also be protected against unauthorized installation of false keys. In specialized equipment this can be achieved by physical locks on a protected interface. The protocol of a "key gun" should also include this kind of protection in addition to its protection against unauthorized reading of keys from false pieces of equipment. In a general software environment there must be a secure way to identify the user before new keys are accepted.

Normally the organization already has some means and rules for the protection of its sensitive data. Often these can be directly applied to the new sensitive objects concerning encryption.

An encryption key should be given at least the same level of protection as would the total amount of data that has or possibly will be enciphered by it.

Organizational matters

An organization that intends to use encryption should be aware of the necessity of continuous efforts to maintain the security level wanted. This has an inevitable effect on the duties of the personnel.

One new specific duty is the careful keeping of a log. The log should clearly show when and why new keys have been installed. It should also contain signatures, verifying that prescribed installation checks have been performed, and other relevant items.

It is strongly advised that the same security organization is used for crypto management as is used for other security matters. As a consequence the security manager or data security manager must learn at least some fundamental facts of encryption and crypto management.

However, other employees at a company must also be taught these matters. Every individual concerned must learn about and be motivated to support the organization's crypto policy as a part of the overall security policy. The more practical duties (key selection, sealing of packets with encryption keys, operating instructions for cryptographic equipment etc) must of course also be taught.

Like all other security devices and systems, crypto systems must be controlled. It may be advantageous if this can be done by the internal audit organization or at least by those who control other security systems.

In this chapter we often make statements like "Physical protection must be provided..." or "It must be impossible for unathorized persons to...". These requirements can not be objectively defined in general terms, and in many circumstances it may also be very difficult to estimate the degree to which they are met. Using an experienced and trusted consultant is in most cases the best solution to this problem.

5 ENCRYPTION AND COMMUNICATION

5.1 Communication principles

The need for standardizing the interconnection between user equipment and networks has always been apparent. This has led to a number of widely accepted recommendations. These specify the interface between the user's data terminal equipment (DTE) and the data circuit terminal equipment (DCE).

For two end systems to communicate with each other there is however also a need for agreed conventions between these. These conventions have been implemented by the different computer vendors mostly as part of their communications software and together they are a part of that vendor's network architecture.

In a world of increasing needs for interconnecting end systems of different makes it has become evident that a comprehensive overall standard for interconnecting end system is of utmost importance. This was the starting point for the ISO work on OSI, Open Systems Interconnection. The basic architectural structure for OSI is a separation in different layers of the various tasks to be performed in the communication between two systems. The reference model, ISO 7498, describes seven layers, where each layer builds on the services provided by the lower layers. For each layer there is a service definition and a protocol for communicating with the corresponding layer in the other end system or with the network. See figure 5.1.

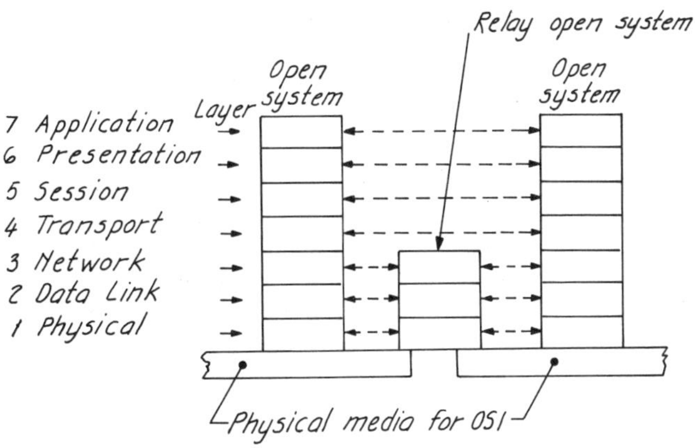

Figure 5.1

The transport layer is the lowest to have end to end significance between systems while the lower layers are network dependent.

Due to the proliferation of proprietary communication systems and the slow process of standardizing and getting acceptance on new protocols it will however take considerable time to achieve the "new world" of universally compatible communication systems. Present protocols will therefore dominate the communication market for years to come, but a slow migration to OSI protocols is foreseen.

The reference model as such has however won widespread acceptance very quickly and today all parties in the data communication support it. This is facilitated by the fact that most present communication architectures already use the layered approach and mapping to the reference model is often possible.

For newly developed architectures the OSI work is adopted as a matter of fact. Examples here are the telematic services from CCITT (teletex, facsimile group 4), the functional standards endorsed by EC in Europe, and MAP (i.e.. Manufacturing Automation Protocol) and TOP (i.e.. Technical Office Protocol) from the U.S.A.

When discussing the implementation of enciphering in the communication process it is therefore natural to relate it to the OSI Reference Model.

5.2 Encryption and the communication architecture

If the encryption process takes place in a high layer of the communication architecture, i.e. close to the application or the production of the text to be ciphered, it may be independent of the type of network used for the communication. In this case the encryption does not have to bother about what line code is used, type of modem, physical interfaces etc. This was the normal way in the earlier days of manual ciphering, when encryption was carried out completely off-line from the communication. With automatic on-line encryption of today's data communication lines the problem is that in order to encrypt in higher layers the encryption process has to interact closely with the corresponding layer(s) of the architecture. These layers are however specific for the different computer suppliers and mostly implemented in system programs and internal hardware. To add on encryption in this environment therefore implies integration with the existing supplier specific products. Physical interfaces for connecting hardware encryption do not exist in these layers. The only practical way to implement encryption in higher layers is therefore to have the data, messages or files, transferred to an application program which carries out the encryption in an off-line mode before transmission takes place. The question then is what to do if the receiver is an unintelligent terminal? Evidently this approach is apt mainly for file encryption for storage protection or communication between computers, e.g. PC's.

In terminal networks encryption in higher layers must probably be implemented and supported by a supplier who controls the entire system including computers and terminals.

Another approach that may be possible for transparent (e.g. circuit switched) networks is to let a crypto device in the DTE-DCE interface react to a specific bit or character string in the user data for starting encryption and another string to stop it. This may be looked upon as an application layer encryption although the connection of crypto devices is made at a low level.

This allows for user selected encryption of specific user data only. It is however not a generally possible solution.

Still another approach, that has been used by some manufacturers, is to have the crypto device connected physically in between DTE and DCE but implement the encryption between the network and transport layer to provide the end-to-end encryption. This means that the crypto device must be able to analyze the data according to the protocols up to layer 3 and submit only the transport data units to enciphering. The ciphered data is then repacked again through layers 3 to 1 and transmitted. This procedure has been tried for packet switching in spite of the difficulties involved.

The international efforts of ISO to standardize on all layers in the communication architecture (OSI) may in due time make it possible to integrate encryption in higher layers with well defined interfaces.

Until then the current implementations of enciphering with stand-alone equipments connected to the separate communication lines will dominate. These equipments work typically in the physical layer (in the terms of the OSI model), though there are also some examples where they interact with the link or network layers. One advantage is that the interfaces to the physical layer have a long story of international standardization and they are mostly available for interconnecting of "extra equipments" like encryption devices. Crypto devices working on the physical layer are certainly the most common on the market today. They have the property of enciphering all data that is passed over the interface including addresses, protocol information and so on. In this aspect they are universally useable for all digital transmissions and they give a good protection against traffic analysis (who is sending to whom, how often, how long messages etc). But the drawback is the need for transparent networks. Intelligent intermediate nodes for switching or storing necessitates deciphering and so new ciphering must be performed on every link that need protection. Leaving the information in clear in the network may be unacceptable in many cases and the required number of crypto devices will soon be prohibitive.

If the crypto devices interact with the link layer protocol a more effective enciphering may be the result. Pure protocol information may then pass in clear. However, this means that these devices normally, but to varying degrees, are communication dependent and that caution must be taken to ensure that the encryption function does not disturb the actual or planned communication environment.

The rest of this chapter will deal with aspects on how encryption may be introduced in different communication situations and some practical implementation problems that may be encountered.

5.3 Practical examples

Leased lines To encrypt leased telephone lines point-to-point is normally trivial, as the line is completely transparent and no network addressing or other information has to be transmitted in cleartext over the interface. Encryption devices connected in the interface between DCE and DTE can thus work in the physical layer and cipher all information passing there. Even if multiplexers are used (for example statistical multiplexers for aggregating several asynchronous connections) transparency is normally kept and the individual connection can easily be encrypted. In this case it should be noted however that substantial savings in crypto devices can be made if the aggregate line is protected with a pair of devices instead of the individual terminal lines. This is possible if the individual lines up to the multiplexer do not need ciphering and it is acceptable to have the same key for all this traffic.

When more intelligent concentrators are used or synchronous lines are multiplexed there may be transparency problems and a closer investigation is needed.

For multipoint lines crypto devices working in the physical layer (i.e. interface signals initiate the ciphering, everything is ciphered) can be used in the same way. Evidently this requires that all drops are equipped with crypto devices and the same key is used for all terminal drops. If this is not acceptable, crypto devices working in the link layer are needed. The host-situated device then must be able to hold different keys for all terminal drops and be able to choose the right one according to the terminal address presented in the message header in the specific link protocol used. In this context it can be noted that presently there are no cost-effective ways to cipher the connection between a clustered terminal and the cluster controller. In some cases this may motivate the use of single stations instead of clustered.

Switched telephone lines. In the public switched telephone network (PSTN) the connection is transparent after call set-up. If this is done manually there are usually no problems to have encryption initiated when data transmission is started. Several new methods for automatic call set-up (with dialling information etc. coming from the DTE via the normal data transmission interface) are however becoming increasingly popular. If one of these are used it is necessary to check if the call set-up commands used can be transported in clear through the intermediary crypto device. This depends on how the different interface signals are used and interpreted in the call set-up procedure respectively by the crypto device.

Circuit switched public data networks, (CSPDN). These networks have the same property of transparency after call set-up as PSTN. This means that the crypto devices must be transparent in the call set-up phase and automatically switch to enciphering when the data transfer phase is established.

Packet switched public data networks, (PSPDN) In packet switching packets are routed between nodes in the network. Every link between nodes has its own line procedure for error correcting. This means that the information received at the node must follow the rules in that procedure and also that some information e.g addresses must be in cleartext in the nodes. Due to the complexity of the internationally standardized packet switching interface (X.25) it is not surprising that crypto devices that can handle this situation on an end-to-end basis have been almost non-existent. (Recently, however, some companies have announced products which offer end-to-end encryption in X.25 networks).

This is clearly a situation where the crypto function should be implemented above the network layer. The difficulties are even more severe if traffic is to be ciphered between an asynchronous terminal accessing the packet network through the PSTN (X.28 interface) and a host with the X.25 interface. There are no stand-alone crypto devices developed for this application so far.

Facsimile. For transmission of documents by facsimile equipment the PSTN is normally used. The modern equipments, belonging to CCITT group 3 or 4, use digital transmission and standardized modems which has made it possible to use normal data enciphering devices for encryption. Still there is a problem due to the common integration of the modem in the facsimile equipment. This means that the interface to interconnect the crypto device may not be available. If the crypto function is not integrated in the facsimile it is necessary to choose equipments where this interface is available.

Telex. Dedicated networks of the PSTN type are normally used for telex traffic. However nationally there may be character strings that are not transparent through the exchanges. Some customers may prefer to have incoming ciphered messages received and stored in ciphered form until an authorised operator arrives and initiates the decryption in a local mode. This case puts heavy restrictions on what characters and character strings can be allowed in cipher texts. For this reason some very special telex crypto devices, which not only follow the code, speed and interface of telex but also make local ciphering and deciphering possible, have been developed. Modern telex equipments have integrated call set-up functions and use key-board dialling. The signalling from the telex equipment to the local exchange may be quite special compared to data networks (e.g. using polarity switchings). So even if common V.21 modems are used instead of the traditional current loops telex cryptos have to be quite special. As a matter of fact it is easier to introduce telex ciphering in an "older" environment with separated call set-up boxes. Special solutions may also be necessary when the newly developed "telex interfaces" for computers, PC's and word processors are used dependent on where the ciphering function is performed.

Teletex. National teletex networks may be implemented in PSTN, CSPDN or PSPDN. As international traffic is of prime importance for teletex users, it is evident that ciphering on a low layer is no real solution as these layers differ too much between the networks. Consequently there are no universal crypto devices available for teletex. Integration in a higher layer in the teletex protocol ought to have been standardized, but this has not happened. It remains for the users to design their own solution in a higher layer, preferably at the document level. This may be possible in computers or PC's that are connected to teletex networks.

Videotex. From the terminal's point of view videotex traffic is composed of asynchronous, low speed characters and should be easy to cipher. In real life there may be quite a few problems to take care of! One of the easier is to take care of the commonly used split speed, which most crypto devices do not offer. More tricky is the fact that the videotex network which mostly uses the PSTN for terminal access often comprises a videotex server of some kind. This server may have features like log-in control, editing, electronic mailboxes and a switching capacity so that the terminal is connected to the computer that stores the sought-after information. The server may be connected to several external computers and these connections may be implemented through leased lines with a synchronous protocol or through a packet-switching network, just to name two common possibilities. It should then be evident that the desired end-to-end protection of the information with ciphering is not so easily implemented. Some traffic to and from the terminal is for the purpose of the server, some information is passing through to the data base in another location.

Evidently we have again a situation where it would be desirable to have encryption placed in higher layers. The alternative, to make the crypto devices follow videotex protocols, is not viable due to the proliferation and instability of these protocols. Consequently dedicated crypto devices for videotex do not exist for universal use.

Interworking networks. As just demonstrated videotex applications may use the concept of interconnected networks. This is always a difficult situation if ciphering is desired from end to end. Other examples are interworking telex/teletex and mailbox systems which can be accessed from different networks. If encipherment can not be implemented in a high enough layer it remains to decipher all traffic before the conversion point and reencipher in the next network. This may be impossible for economical, practical or security reasons.

Line protocols. For the vendors of crypto devices it is naturally interesting to have them independent of the protocols used in the link layer as this broadens the possible market. It is also an advantage many times that addresses and other protocol information are protected. But as we have seen there are situations where conforming to protocol and code is necessary, viz when there is some protocol dependent equipment intervening between the endpoints. But there may be other reasons for having the crypto function designed for special communication protocols. One is the inherent overhead introduced by the cryptological synchronization at every turn-around of a halfduplex line. Though small, it may be undesirable in a polled environment, where every poll and acknowledgement initiates a new cipher start. A protocol sensitive crypto device can leave these procedural elements unencrypted and only encrypt user data. Another reason may be that for cryptological reasons the continous encipherment of stereotype information like protocol headers and idle signalling ought to be avoided.

Physical interfaces. The most common interface, CCITT V.24, is known to have caused many problems when intermediate crypto devices are installed. This is primarily due to the flexibility and many options that are possible with the interchange circuits defined in V.24. Sometimes expected interchange circuits are not implemented or they are not used in the expected relation to other circuits. One example is that many crypto devices were designed for halfduplex and/or switched traffic and they naturally use the interchange circuits 105 (request for sending) and 106 (ready for sending) to initiate crypto synchronization. If installed on a leased line these interchange circuits may not be used or be constantly on which may cause problems with crypto synchronization. Other similar problems may be encountered with respect to interchange circuit behavior when modems with automatic equalizing are used or when automatic call set-up procedures (e.g. CCITT V.25) are used. These problems can sometimes be solved by strapping or parameter setting in DCE, DTE and/or crypto device. As mentioned, the interfaces in telex and facsimile equipments may carry call set-up commands. These have to be passed without alteration through the crypto device. An investigation on how the proposed crypto equipment can handle such situations is necessary. A test period is always recommended before a decision is taken on the purchase of specific crypto hardware!

6 STANDARDS FOR DATA SECURITY

Development of standards for data security began within the domain of the United States government with the publication of the well-known Data Encryption Standard (DES), a block encipherment algorithm, in 1977. Since that time other data security related standards have been published or are in preparation by other bodies. The International Standards Organisation (ISO) began work on this subject in 1980 with the establishment of Working Group 1 (WG1) of Technical Committee 97 (TC97) (the technical committee responsible for information processing); WG1 later became a sub-committee (SC) of TC97 under the designation of TC97/SC20.

Data security standards fall under several headings - algorithms, modes of operation, enhancement of communication protocols, key management, authentication, etc. The DES, a complete definition of an encipherment algorithm, was re-written in international terms, whilst retaining the same internal logical structure, and a lot of work was done in preparation for its publication as an international standard; this has been known as Data Encipherment Algorithm 1 (DEA1), with the ISO designation 8227. This text was functionally identical with the US DES, though there were minor differences relating particularly to implementation of the algorithm in software and the question of parity bits in the keys; this text was nearly ready for publication as a standard. Also in preparation has been a standard defining the RSA public key cryptosystem. The latter was to be a brief statement of the RSA algorithm, with some advice appended on implementation and choice of keys. The ISO designation was 9307; this had progressed to 'Draft Proposal' status.

At this point it is necessary to say that the preparation for publication of international standards for data encipherment has been discontinued. At its meeting in May 1986 TC97 made a reference to ISO Council to decide whether it was right that encipherment should be subject to standardisation; one major concern was that international standardisation of the DES might encourage overdependence on one algorithm, with consequential greater risks should that algorithm be broken. The decision taken by ISO Council is that DEA1 shall not be published as an international standard; this decision has recently been extended to cover all algorithms, including the RSA. Thus there will be no international standards for data encipherment algorithms published under the auspices of ISO TC97. A consequence of this decision is the establishment by ISO of a register of encipherment algorithms; the intention of the establishment of the register is to offer choice to users of encipherment systems, thereby reducing the sensitivity of any particular algorithm on the register as a potential target for criminal attackers. The decision not to proceed to publication of a standard for DEA1 is not likely to discourage use of the DES algorithm in the short term, especially in the financial world in which it has proved so popular. Indeed the DEA1 algorithm is very likely to appear on the ISO Register when the latter is created. Encouragement will be given for the inclusion of other algorithms in the register, including some which will not be published. An entry in the register may therefore be no more than a definition of the external appearance (block size, key size, etc) of the algorithm, though other algorithms including the DEA1 may be fully specified.

Other standards for data security are not subject to the same concerns as those related to encipherment algorithms, so progress is continuing in most of the other relevant areas. In using a block cipher algorithm it is necessary to specify ways in which it can safely be applied to units of text larger than the block; the object is to prevent undetected interference with a stream of enciphered data units. Secure ways of using a block cipher have become known as 'modes of operation'. Modes of operation for 64-bit block ciphers are covered by ISO Draft International Standard (DIS) 8372, which was written in a sufficiently general way to refer not only to DEA1 but to other ciphers with 64-bit blocks as well. Unfortunately, the references made in DIS 8372 to DEA1 necessitated changes, so this draft has been revised to exclude such references, which were not, in any case, an essential feature of the text; this standard is expected to be issued before the end of 1987. It is proposed that modes of operation for n-bit block ciphers shall be defined in a later standard.

Regarding enhancement of communication protocols, DIS 9160 defines the way in which encipherment can be introduced at the physical layer of the Open Systems Interconnection (OSI) communication protocol architecture. Because of specific reference in this standard to DEA1, progress on this standard has also been held up, but it is now likely to go ahead with appropriate changes. It is worth noting that a standard of the American National Standards Institute (ANSI) defines encipherment enhancement of communication protocols at both the physical and link layers; this is X3.105-1983, which specifies the DES as the recommended algorithm.

Data security will, of course, be required in many layers of Open Systems Interconnection. To impose order on what might otherwise become chaotic, an addendum on data security to the OSI definition (ISO 7498) is in an advanced state of preparation (this has now been given the status of a second part of the standard rather than an addendum). This gives advice on the location of various security services and mechanisms such as encipherment, authentication of text and of communicating entities, digital signature, etc. An objective of the addendum is to recommend a limited set of locations for these facilities within the architecture, in order to encourage interoperability. It is perhaps premature to indicate at what layers of OSI we may expect to find certain specific security services and mechanisms, because discussion is still proceeding. However, it seems at present that no such services will be located at the session layer (5) and that provision of security facilities at the link layer (2) will be a temporary expedient, with nothing being provided here in the long term. A section of the draft addendum is devoted to management issues which can include many aspects of controlling the operation of data security services, including the requirement for the provision of certain minimum protection measures and the availability of encipherment keys where these are required. Work is beginning within SC20 to write standards for security enhancements of communication protocols at the network, transport and presentation layers. Other work includes protocols designed to facilitate peer entity authentication. These protocols and protocol enhancements will need to fit into the structure of the security architecture mentioned above.

A particular management issue which has received some attention is that of control of keys for data encipherment. ANSI has published a standard for key management for financial institutions for wholesale applications, ie for inter-bank value transfers; this is X9.17-1985. The latter has been taken as the basis for ISO work which comes under TC68, the banking standards technical committee. With appropriate changes to the text for international use, the current ISO document is DP 8732. It is understood that ANSI is also preparing a standard for retail key management (X9.24), ie for the protection of electronic funds transfer transactions between individual customers and their banks.

Issues that figure prominently in key management concern who shall choose encipherment keys - users or service providers, how they shall be distributed - by courier or by communication channel, how they shall be protected - by master keys or by physical protection (or both), etc. These matters are addressed in such documents as DP 8732, but much work remains to be done.

Authentication of messages is another important issue in data security, providing a means of determining whether a message (which may or may not be enciphered) has arrived at its destination unaltered since its original transmission. This is covered by an ANSI standard X9.9-1984 (a revision of a standard originally issued in 1982) which applies the DES algorithm to achieving data authentication in wholesale transactions. This standard has also been taken in TC68 as the basis for an ISO document under the designation of DIS 8730. Though this refers to the same algorithm (DEA1) as DIS 8227, it is understood that progress on this standard is not currently suspended. ANSI is currently preparing a standard (X9.19) for message authentication in the retail context.

Yet another standard in prepartion in the area of ISO TC68 is one for algorithms for authentication (as distinct from the algebra of the authentication calculation itself); this is designated 8731 and covers a statement of the DES and of two other algorithms. The current status of this document is not known to the writer.

There is little doubt that the area of data security standards where least has been done and most remains to be done is that of security management. However, management must be tailored to reflect the nature of the services and mechanisms that are being controlled. It is therefore, perhaps, inevitable that development of techniques for security management must follow the standardisation of specific methods of achieving the various security requirements. The latter should, however, be designed with efficient and secure management as one of the design objectives.

7 ENCRYPTION IMPLEMENTATIONS

7.1 Hardware

Data encryption products today usually come in the form of hardware. There are three main reasons for this: speed, security, and ease of installation.

Hardware is desirable for speed since most encryption algorithms contain parts which are not suited to the instruction set of a normal general purpose computer. One example is the DES, which contains permutations among 64 bits. Usually no basic computer instruction operates on single bits. Thus this part must be implemented with the help of single bit masking and treatment bit by bit, which is slow, or with the help of huge tables, which cannot be accommodated within a reasonable space. Thus software implementations of the DES usually use one thousand times the time needed for a hardware implementation in a dedicated circuit. Actually the DES is standardized as hardware by NBS, while the ANSI version DEA acknowledges software as well.

Another wellknown algorithm, RSA, suffers from the same problems with speed in implementations as software on a general purpose computer. Software encryption of a single 512 bit block may take as much as several seconds. This is of course not acceptable in a computer environment. Thus commercial RSA implementations today use special design VLSI chips.

In general, encryption algorithms are designed according to their own criteria, not with a view to computer instructions. Thus most algorithms will be faster on dedicated hardware than as software routines. Today hardware can manage speeds up to 2 Mbit/s, and standard equipment is made to fit todays standard transmission speeds of 1.2 to 9.6 kbit/s.

Another aspect on the problem with speed is that encryption tends to keep the processor busy for a considerable length of time every time it is used. Thus it may be worthwhile to implement the encryption as a specific device, even if that device uses a standard processor. Much time is gained anyway due to the parallellism that is introduced.

As for security, the problem is that a general computer contains no physical protection for its contents. Usually the case can be opened, bugs and connections to analysers can be introduced, the contents of storage cells can be revealed etc. A dedicated encryption chip can be securely encapsulated, and the internal logic can be designed so that unencrypted keys remain within the device; there is no instruction for reading them from the outside. The same is true for discrete component implementations of encryption in protected environments.

Another security problem is that electro-magnetic radiation may reveal what is going on inside any piece of electronic equipment. Bugs placed in the vicinity of a device performing encryption may reveal the key to a skilled opponent. This problem can also be taken care of in dedicated hardware.

Finally the ease of installation depends on where you can find standard logical interfaces for your encryption. Today encryption is often intended for communications, at a level in the OSI standard where the protection is not at the discretion of the user and should not hamper his work. Thus if you want to use software it must be embedded deep into the system. This demands good knowledge of the system. No standards exist, and even the same version number of the same operating system may differ in crucial aspects from system to system due to different configurations.

Physical interfaces at this level are, however, standardised and conveniently available. Even the most non-technical user can unplug his terminal, plug a "black box" line into the same place, and finally plug the line from the modem into the "black box". There may be unexpected problems with this approach, as is described in chapter 5. Similar problems may, however, appear also with a software implementation.

Today two kinds of hardware products dominate the market: boxes meant to plug in on a communications line and boards for Personal Computers, PC:s. You can also find encryption as a peripheral, in a smart card etc. These are usually dedicated solutions for special purposes.

Crypto boxes have been on the market for a very long time. Originally the only available equipment types were intended for military use. This was mirrored both in the level of protection and the price. Today both security and price differ depending on the intended use. The market is changing rapidly, and although some stable suppliers have remained in the field for a very long time, a list of current suppliers will in general be invalid within months. A list of products will have an even shorter life. The trend for the moment is, quite naturally, towards higher attainable bit rates and greater flexibility in use. Internal key management is also often of such a quality that user intervention is seldom needed. There are still, however, less advanced products on the market with just one key available. They can be well suited to some situations, but the buyer must always ensure himself about what key management is needed and thus what he himself must provide for any specific type of equipment. The same is true for physical security around the device.

Boards for PC:s are usually intended for encryption of files, which can then be stored or transmitted. (Lack of transparency in the protocol may, however, hinder the transmission of such encrypted files.) Boards are normally meant to operate in a secure environment, which means that circuits are not encapsulated against physical interference, radio-magnetic emanations are not controlled etc. Keys are also usually entered through the PC without any control that they are properly erased afterwards. Thus, PC-boards are often close to software in security, but their main advantage is in speed.

Other solutions are often closely linked to some specific use. One example is the French "smart card" which contains a one-way encryption algorithm. This algorithm is used within the card for some specific applications, but the card complete with algorithm can very well be used as a component in another system where the existence of the protected algorithm can be useful.

Encryption equipment can be designed either to encrypt everything that leaves the computer and decrypt everything that enters, or to encrypt and decrypt the actual message according to a standard communications protocol. This type of equipment can be installed without any additions or changes in the software. In other cases, however, there is a need for driver routines, calls of these routines, routines for finding and possibly marking what should be encrypted etc. Thus installation of hardware for encryption may mean substantial additions and changes in the software as well.

7.2 Software.

Every encryption algorithm can in principle be implemented as a software routine. The disadvantages with this have been mentioned above. The advantages lie in portability and flexibility. A program written in a standard high level language can be installed on any machine supporting that language, which often means most existing computers. Usually a skilled programmer can also convert the program to another language, should that be desirable. The program can then be called whenever the user wants extra protection for some set of data. It can also be included as a part of a larger system, for example a general protection package or a communications package.

Software encryption programs are today rather popular as protection for PC files. Such encryption is meant to protect files when they are sent to other locations or when the PC is left unattended and unprotected. It is then important that the user gets a key management scheme which suits his needs. It is also important that the program erases keys and unencrypted files after the encryption. Important data should never be left by the program on secondary media such as in work storage on a whinchester disk. Erased files must be filled in with nonsense data, not just erased as an entry in the index file. If the encryption program does not itself erase keys and other sensitive data in primary storage, the user should switch off the PC and restart it after a completed encryption session or at least before leaving the room. Otherwise these sensitive data can be read by a skilled person with access to the PC.

7.3 Choosing an implementation

Obviously there is no encryption product that is ideal for all users and all levels of protection. Some differences have been treated in the preceding sections. It is of course up to the buyer to decide what he really needs. The important thing is that a few questions must be answered:

Why should I consider encrypting data at all? This has been treated in the introduction.

What kinds of threats against data can be eliminated by cryptographic protection? This has been treated in chapter 2.

What does it mean to a buyer that a certain product uses a certain algorithm? This has been treated in chapter 3.

How is the cryptographic protection protected? What is provided in the product and what must the user add? This is treated in chapter 4.

How will the added cryptographic protection fit into (and possibly upset) existing routines and protocols? This is treated in chapter 5.

Are there any standards in the field? This is treated in chapter 6.

What kinds of products are there on the market? This is treated in this chapter.

8 GLOSSARY AND ABBREVIATIONS

Active wiretapping
An attempt to change the flow of data over a communications line by deleting, inserting or altering data.

Algorithm
A set of formulas and rules describing the way a computation or a task is to be performed.

ANSI
American National Standards Institute

Asymmetric encryption
All cryptographic methods where the decryption key is not the same as the encryption key and the decryption key can not be calculated from the corresponding encryption key. See Public key cryptography.

Authentication
a) Verification of the identity of an entity, such as a person or a piece of equipment
b) A method to ensure that a message originates from the stated source, that it is not altered and that it is not a repetition of an earlier genuine message.

Authenticator
A data item added to a message in order to make it possible to perform authentication.

Block
A sequence of bits, usually of a fixed length, which is treated as a separate entity.

Block cipher
A cipher method which acts on a block of fixed length and where the resulting cryptogram is defined completely by that block and the key. Thus the result of a block cipher depends neither on the contents of previous blocks nor the position of the current block in a sequence of blocks.

Brute force attack
Cryptanalysis by using exhaustive search

CBC
Cipher Block Chaining

CCITT
Comité Consultatif International Télégraphique et Télephonique

CFB
Cipher Feedback Mode

Chaining
A technique to make the encrypted value of one block depend on previous blocks in a chain when a block cipher algorithm is used.

Cipher
A cryptographic method using an algorithm as opposed to using a code table.

Ciphertext
The enciphered form of data such as texts, figures, messages etc; the result of enciphering.

Code
A cryptographic method in which a table is used to translate cleartext into cryptogram. (Note that a block cipher with a fixed key can be expressed as a translation table and thus as a code!)

Complexity
A measure of the degree of difficulty of a calculation. Complexity is usually expressed in terms of the number of elementary operations required. When scientific stringency is not necessary complexity is often expressed as the time required to do the calculation on a given computer.

Computer virus
A trojan horse program part which copies itself into other programs.

Cryptanalysis
Finding the plaintext behind a cryptogram without prior knowledge of the key, finding the key, finding the algorithm used, or finding any combination of these.

Cryptogram
The encrypted form of data; the result of encryption.

Cryptography
The technique of concealing the content of a message by a code or a cipher. Cryptographic techniques are also used for other purposes such as authentication.

Cryptographic Check Value
A cryptographic message authenticator.

Cryptographic synchronization
The sending of an IV, a message key or any such data entity in order to give pseudorandom number generators, block chaining etc a common starting value at both ends of an encrypted connection.

Cryptology
The science treating cryptography and cryptanalysis.

Decipherment
The reversal of encipherment, recreating the plaintext

Decoding
The reversal of encoding; recreating the plaintext.

Decryption
The reversal of encryption. Often used as a synonym of decipherment.

DEA
An ANSI encryption standard based on the DES.

DEA-1
Proposed and subsequently withdrawn ISO standard based on DEA.

DEA-2
Proposed and subsequently withdrawn ISO standard based on RSA

DES, Data Encryption Standard
A US Federal standard for encryption, which has been widely used in all kinds of non-military environments, especially in financial institutions.

Digest
A short bit sequence created from a longer bit sequence such as a message. The value of a digest must depend on all the bits in the longer sequence.

Digital signature
The result of a public key transformation depending on all the bits of a message and on a secret key. The authenticity of the message and its originator can be verified by anyone using the associated public key.

ECB
Electronic Codebook Mode

Encipherment
Creating a cryptogram by applying a cipher algorithm to a plaintext.

Encoding
Creating a cryptogram by applying a code to a plaintext.

Encryption
Creating a cryptogram by applying a method of cryptography to a plaintext. Often used as a synonym to encipherment.

End-to-end encryption
A technique whereby transmitted messages are encrypted at their point of origin and not decrypted until they reach their final destination.

Exclusive OR
An operation on bits usually designated by ≈, where 0 ≈ 0=0, 0 ≈ 1=1, and 1 ≈ 1=0

Exhaustive key search
Solving the problem of finding the correct value for a key by testing possible values until the correct one is found.

Hash function
A data compression function. Hash functions with special one-way properties are used for cryptographic purposes.

IV, Initializing Value, Variable or Vector
The starting value for chaining information for a block cipher used with chaining or the starting value for a pseudorandom number generator used for a stream cipher.

Key
A sequence of symbols controlling encipherment and decipherment. The security of a cryptogram depends on the secrecy of the decipherment key.

Key generator
A piece of equipment generating keys with desired random properties.

Key gun
A secure electronic transport device for keys.

Key space
The range of values which a key may take. A large key space is necessary to prevent exhaustive search for a key.

Link encryption
A technique where a transmitted message is encrypted only while it is passed over a link in a network. Messages appear as plaintext in nodes. Link encryption in a network may be restricted to specific links, leaving messages unprotected as they pass other links.

MAC, Message Authentication Code
A name for an authenticator (thus not a code in the cryptographic sense).

MDC, Manipulation Detection Code, Modification Detection Code
A data field appended to the plaintext in order to make it possible to detect illegal changes to the ciphertext (again not a code in the cryptographic sense).

Message key
See IV. Message key is also sometimes used for any key encrypting just one message.

Mod
Short for modulo.

Modulo
In calculations modulo some integer n, the result is always smaller than n. This is achieved by subtracting n from any intermediate value greater than or equal to n repeatedly until that value is smaller than n. This means that addition modulo 2 is equivalent to exclusive OR.

One-way function
A function which is relatively easy to compute and the inverse of which is in practice exceedingly difficult to compute. Thus for a one-way function there exists an algorithm which will within reasonable time find the value of the function for any given input, while there is no known way to find that input which will result in a given value for the function.

PIN, Personal Identification Number
A password using only decimal digits, which may be linked to a card, such as a bank card for automatic teller machines. A PIN is often linked to the card number through a cryptographic function.

Public key cryptography
An asymmetric ciphering technique using specific one-way properties of some mathematical functions in such a way that it is possible to make the encryption key public without endangering the secrecy of the decryption key.

RSA
A public key encryption algorithm named after its inventors Rivest, Shamir and Adleman.

Scrambling
Rearranging the elements of a sequence in a prescribed way in order to make the transmitted sequence unintelligible to intruders. Often used to protect analog transmissions from eavesdropping.

Semi-weak keys
See weak keys.

Stream cipher
A cipher designed to deal with messages of any length. Usually this is achieved by using the key as a seed for an algorithm, which creates an infinite stream of random looking bits, which are then added to the plaintext. In a stream cipher the cryptogram form of a specific part of the plaintext will always depend on the position of that part within the plaintext.

Symmetric encryption
All cryptographic methods where the key for encryption is the same as the key for decryption or the encryption key can be easily calculated from the decryption key.

Traffic analysis
Detection of who is sending messages to whom and in what quantities in a given network.

Trojan horse
Executable computer code existing in a program unknown to the user.

Virus
See computer virus

Weak key
A value for an encryption key which may facilitate cryptanalysis of messages encrypted with that key or endanger the security of an authenticator or a digital signature. The property of weakness is always tied to a specific algorithm. Some sets of keys, which may be regarded as not so dangerous to use, are sometimes called semi-weak keys.

Wiretapping
Recording or just listening to any signals passing on a transmission line. Wiretapping without attempts to alter or insert signals is sometimes called passive wiretapping. See also active wiretapping.

XOR
See Exclusive OR

9 REFERENCES

1. Basic books

Beker H and Piper F: "Cipher systems: The protection of Communications", Northwood Books 1982.

Davies D and Price W: "Security for Computer Networks", John Wiley & Sons 1984.

Denning D: "Cryptography and Data Security", Addison Wesley 1982.

Meyer C and Matyas S: "Cryptography. A New Dimension in Computer Data Security", John Wiley & Sons 1982.

2. On where and how to use cryptographic protection

Albert D and Morse S: "Combatting Software Piracy by Encryption and Key Management", IEEE Computer, Vol 17, No 4, April 1984, pp 68-73.

Blakley G R, Meadows C and Purdy G B: "Fingerprinting long forgiving messages", Advances in Cryptology, Proc. CRYPTO'85, Springer Verlag LNCS 218, pp 180-191.

Blum M, Vazirani U and Vazirani V: "Reducibility Among Protocols", Advances in Cryptology, Proc. CRYPTO'83, Plenum Press, 1983, pp 137-146.

Branstad D. "Considerations for security in the OSI architecture", IEEE Network Magazine, Vol 1, No 2, April 1987, pp 34-39.

Chaum D: "Showing Credentials Without Identification, Signatures Transferred Between Unconditionally Unlinkable Pseudonyms", Advances in Cryptology, Proc. EUROCRYPT'85, LNCS 219 Springer Verlag, 1986, pp 241-244.

Chaum D: "Security Without Identification: Transaction Systems to Make Big Brother Obsolete", Communications of the ACM, Vol 28, No 10, October 1985, pp 1030-1044.

Cohen F: "Algorithmic authentication of identification", Information Age, Vol 7, No 1, Jan 1985, pp 35-41.

Cohen F: "A Secure Computer Network Design", Computers & Security, Vol 4, No 3, September 1985, pp 189-205.

Cole G: "Design alternatives for computer network security", National Bureau of Standards Special Publication 500-21, Vol 1, January 1978.

Davida G, Wells D and Kam J: "A Database Encryption System with Subkeys", ACM Transactions on Database System (TODS), Vol 6, No 2, June 1981, pp 312-328.

Davies D: "Using the signature token to create a negotiable document", Advances in Cryptology, Proc. Crypto'83, ed. D. Chaum, Plenum Press 1984, pp 377-382.

De Millo R and Michael M: "Protocols for Data Security", IEEE Computer, Vol 16, No 2, Feb 1983, pp 39-54.

Even S, Goldreich O and Lempel A: "A Randomized Protocol for Signing Contracts", Communications of the ACM, Vol 28, No 6, June 1985, pp 637-647.

Fiat A and Shamir A: "How to Prove Yourself, Practical Solutions to Identification and Signature Problems", Advances in Cryptology, Proc. CRYPTO'86, LNCS 263 Springer-Verlag, 1987, pp 186-194.

Goldreich O: "Towards a Theory of Software Protection", Advances in Cryptology, Proc. CRYPTO'86, LNCS 263 Springer Verlag, 1987, pp 426-439.

Gosler, J R: "Software protection: Myth or Reality?" Advances in Cryptology, Proc. CRYPTO'85, Springer Verlag LNCS 218, pp 140-157.

Guillou L and Lorig B: "Cryptography and teleinformatics", Computers & Security, Vol 1, No 1, January 1982, pp 27-33.

Herzberg, A and Pinter, S: "Public protection of software", Advances in Cryptology, Proc. CRYPTO'85, Springer Verlag LNCS 218, pp 158-179.

Highland H J: "Microcomputer Security: Data protection techniques", Computers & Security, Vol 4, No 2, June 1985, pp 123-134.

Highland H J: "How to Evaluate Microcomputer Encryption Software and Hardware", Computers & Security, Vol 6, No 3, June 1987, pp 229-244.

Holloway, C J: "The concept underlying security in an EFT/POS network", Proc. European Computer Systems Security Forum, Online, London, November 1984, pp 31-40.

Jueneman R, Matyas S and Meyer C: "Message Authentication with manipulation detection codes", Proc. 1983 Symposium on Security and Privacy, Oakland CA, April 1983, pp 33-54.

Jueneman R: "Electronic document authentication", IEEE Network Magazine, Vol 1, No 2, April 1987, pp 17-23.

Krause L. "Data encryption in ISO", Computers & Security, Vol 3, No 3, August 1984, pp 234-236.

Maude T and Maude D: "Hardware protection against software piracy", Comm. ACM, Vol 27, No 9, Sept. 1984, pp 950-959.

Nestman C, Windsor J and Hinson M: "Tutorial on telecommunications and security", Computers & Security, Vol 3, No 3, August 1983, pp 215-224.

Purdy G B, Simmons G J and Studier J A: "A software protection scheme", Proc. of the 1982 Symposium on Security and Privacy, Oakland CA, April 1982, pp 99-103.

Schaumüller I and Piller E: "A method of software protection based on the use of smart cards and cryptographic techniques", Advances in Cryptology, Proc. EUROCRYPT'84, Springer Verlag LNCS 209, pp 446-456.

Wood C: "Floppy Diskette Security Measures", Computers & Security, Vo 14, No3, Sept 1985, pp 223-228.

3. Key Management

Beker H, Friend J and Halliden P: "Simplifying key management in Electronic Funds Transfer Point of sales systems", Electronics Letters, Vol 19, No 12, June 1983, pp 442-444.

Beker, H J. "Key management issues: a simple protocol for 'insecure' terminals", Proc. of Int. Conf. on System Security, System Security - the Technical Challenge, Online, London, October 1985, pp 245-263.

Denning D: "Protecting public keys and signature keys", IEEE Computer, Vol 16, No 2, Feb 1983, pp 27-35.

Ehrsam S, Matyas S, Meyer C and Tuchman W: "A cryptographic key management scheme for implementing the Data Encryption Standard", IBM Systems Journal, Vol 17, No 2, 1978, pp 106-125.

Jansen C: "On the key storage requirements for secure terminals", Computers & Security, Vol 5, No 2, June 1986, pp 145-149.

Jones R: "Some techniques for handling encipherment keys" ICL Technical Journal, Nov 1982, pp 175-188.

Kolata, G B: "Flaws found in popular code", Science, 219, January 28, 1983, pp 369-370.

Matyas S and Meyer C: "Generation, distribution and installation of cryptographic keys", IBM Systems Journal, Vol 17, No 2, 1978, pp 126-137.

Newman D, Omura J and Pickholtz R: "Public key management for network security", IEEE Network Magazine, Vol 1, No 2, April 1987, pp 11-16.

Pagels-Fick G: "Implementation issues for master key distribution and protected keyload procedures", Computers & Security: A Global Challenge, Proc. IFIP SEC'84, (ed Finch and Dougall), North-Holland 1984, pp 381-390.

Price W: "Key management for data encipherment", Security: Proc. of IFIP/SEC'83, (ed. V. Fåk), North Holland, 1983.

Smid, M E: "A key notarization system for computer networks", US Dept. of Commerce, NBS Special Report 500-54, October 1979.

4. Public key systems and digital signatures

Akl S: "Digital Signatures: A Tutorial Survey", IEEE Computer, Vol 16, No 2, Feb 1983, pp 15-26.

Davies D: "Applying the RSA Digital Signature to Electronic Mail", IEEE Computer, Vol 16, No 2, Feb 1983, pp 55-62.

Davies D: "Public key ciphers and signatures", Information Age, vol 6, No 1, Jan 1984, pp 25-29.

Denning D: "Digital signatures with RSA and other public key cryptosystems", Comm. of the ACM, Vol 27, No 4, April 1984, pp 388-392.

Diffie W and Hellman M: "New directions in cryptography", IEEE Trans. on Information Theory, Vol 22, No 6, Now 1976, pp 644-654.

Dolev D and Yao A: "On the security of public key protocols", IEEE Trans. on Information Theory, vol 29, No 2, March 1983, pp 198-208.

Gordon J: "Strong RSA keys" Electronics Letters, Vol 2, No 12, June 1984, pp 514-516.

5. Equipment

Müller-Schloer, C: "A Microprocessor based Cryptoprocessor", IEEE Micro, No 10, Oct 1983, pp 5-15.

Serpell S: "Cryptographic equipment security: A code of practice", Computers & Security, Vol 4, No 1, March 1985, pp 47-64.

6. On the strength of algorithms

Goeltz T: "Why not DES", Computers & Security, Vol 5, No 1, March 1986, pp 24-27.

Diffie W and Hellman M: "Exhaustive cryptanalysis of the NBS Data Encryption Standard", Computer, June 1977, pp 74-84.

Kochanski M: "Data Insecurity Packages", Cryptologia, Vol 11, No 1, Jan 1987, pp 1-15.

van den Assem R and van Eik W J: "A Chosen Plaintext Attack on the Microsoft Basic Protection", Computers & Security, Vol 15, No 1, March 1986, pp 36-45.

7. Standards

National Security Agency:

Federal Standard 1027, 1982: "Telecommunications: general security requirements for equipement using the Data Encryption Standard"

National Bureau of Standards:

FIPS publ 46, 1977: "Data Encryption Standard"

FIPS publ 81, 1980: "DES Modes of Operation"

FIPS publ 74, 1981: "Guidelines for implementing and using the NBS Data Encryption Standard"

ANSI:

X3.92, 1981: "Data Encryption Algorithm"

X3.106, 1983: "DEA - Modes of Operation"

X3.105, 1983: "Data Link Encryption"

X9.8, 1982: "Personal Identification Number (PIN) Management and Security"

X9.9, 1984: "Financial Institution Message Authentication"

X9.17, 1985: "Financial Institution Key Management (Wholesale)

ISO 7498, 1984: "Information Processing Systems - Open Systems Interconnection - Basic Reference Model"

INDEX